D1582750

This book is due for return on or before the last date show

The Institute of Biology's
Studies in Biology no. 42

The Structure and Function of Enzymes

by *Colin H. Wynn* Ph.D.,
Senior Lecturer in Biological Chemistry,
University of Manchester

Edward Arnold

© Colin H. Wynn 1973

First published 1973
by Edward Arnold (Publishers) Limited,
25 Hill Street,
London, W1X 8LL

Boards edition ISBN: 0 7131 2401 6
Paper edition   ISBN: 0 7131 2402 4

Printed in Great Britain by
The Camelot Press Ltd, London and Southampton

General Preface to the Series

It is no longer possible for one textbook to cover the whole field of Biology and to remain sufficiently up to date. At the same time students at school, and indeed those in their first year at universities, must be contemporary in their biological outlook and know where the most important developments are taking place.

The Biological Education Committee, set up jointly by the Royal Society and the Institute of Biology, is sponsoring, therefore, the production of a series of booklets dealing with limited biological topics in which recent progress has been most rapid and important.

A feature of the series is that the booklets indicate as clearly as possible the methods that have been employed in elucidating the problems with which they deal. Wherever appropriate there are suggestions for practical work for the student. To ensure that each booklet is kept up to date, comments and questions about the contents may be sent to the author or the Institute.

1973

INSTITUTE OF BIOLOGY
41 Queen's Gate
London, S.W.7

Preface

The ambition to explain all biological phenomena in chemical and physical terms is no unreasonable one. This booklet is intended to place the nature of the reactions taking place and the catalysis of those reactions on such a basis. Several enzymes have been considered in detail and an attempt made to correlate the structure of the enzyme with the catalysis mechanism. Although our understanding of the mechanism of enzyme catalysis has increased recently and the correlation of structure and function is feasible in some systems, it is hoped that this booklet will provide the stimulus for others to take up the challenge. The ultimate aim of any such work must be the complete chemical description of the cell.

Manchester
1973

C. H. W.

Contents

Enzymes as Chemical Catalysts   1

1.1 The need for catalysis

A common feature of the myriad chemical reactions that take place in a living cell is that these reactions occur at a far faster rate than similar reactions occurring in a non-living environment. An outstanding example of this is the contraction of a muscle where tremendous energy is used and must be made available in a very short period of time. To account for this phenomenon of the increase in reaction rate it is necessary to postulate the existence of catalysts. Catalysts find widespread use in many industrial processes and their properties have been studied extensively. In the specific case of catalysts of biological origin the term 'enzyme' is used, meaning in yeast, and showing historically that much of our knowledge of the properties of enzymes was derived from studies on yeasts and other microorganisms. Thus enzymes may be defined as molecules of biological origin which increase the rate of specific reactions, although not affecting the final position of the equilibrium established, and which may be recovered from the reaction mixture at the end of the reaction.

Another essential feature of enzymes is that their concentration will also affect the rate of reaction and so we can consider that, by varying the concentration of enzyme, subtle control of the extent of metabolic reactions may be made. In addition to this control mediated by the concentration of enzymes, the catalytic activity of the enzymes themselves is also subject to control. In this way although the concentration of enzyme may be constant, the ability of the enzyme to influence a particular reaction may vary. Although the enzyme composition of the muscle is essentially the same during contraction and relaxation, there are very different reactions proceeding during these two phases of muscular activity.

Since earliest recorded time man has made use of the properties of enzymes in the production of alcohol and in cheese-making. At first this use must have been purely accidental and ill-controlled, but with the passing of time it is probable that early man began to realize the influence of warmth and of the other constituents on these processes called fermentation and took unwittingly the first steps in biochemistry. Slow progress ensued and it was not until the late eighteenth and early nineteenth century that more detailed investigations of the processes of fermentation were undertaken. Among the major contributions of this period were those of Schwann who recognized that yeast was a plant capable of converting sugar to alcohol and carbon dioxide and of Pasteur who studied the influence of oxygen on these processes and analysed the end product of many different fermentations. Unfortunately further progress at this time was impeded by a widespread belief in vitalism, which held that a vital force was necessary for the synthesis of organic compounds and that this vital force was present only in living organisms. Finally Buchner's demonstration of fermentation of sugar by a yeast extract from which all living yeast had been removed paved the way for the detailed examination of the chemical and physical properties of the enzymes.

In the early twentieth century the individual enzymes involved in fermentation were isolated and the intermediate compounds formed in the production of alcohol from sugar were found. The last twenty years have seen tremendous advances in our knowledge of both the pathways of metabolism and the enzymes responsible. The necessity for control and the way in which it is brought about have also been clarified. These advances have been made possible by the development of modern technology and by the rejection of the concepts of vitalism. The belief that all the processes that occur within a living cell are capable of interpretation by the laws of physical science has been rewarded by the rapid advances that have been made; a belief in vitalism inhibits experimentation.

It is constructive to consider for a given reaction the mechanism of the uncatalysed reaction and to compare this with what is known about the mechanism of the catalysed reaction. In this way we shall be able to realize the true nature of catalysis and its control.

1.2   Ribonucleic acid and its hydrolysis by acid and alkali

Ribonucleic acid, commonly abbreviated as RNA, is a macromolecule of biological importance in the growth of cells, since, amongst its many functions is that of carrying the genetic information contained in the nucleus to the cytoplasm where it is translated into specific proteins. The essential features of a segment of the RNA molecule are shown in Fig. 1–1 and the main features of the structures summarized.

When RNA is treated under fairly stringent conditions with acid or alkali, the molecule is completely hydrolysed yielding mixtures containing molar proportions of phosphate, ribose and heterocyclic bases. However, when RNA is hydrolysed under less drastic conditions, for example 0·1M sodium hydroxide at room temperature, various larger fragments can be identified. It is interesting to consider in detail the mechanism of action of alkali on RNA. The first stage in this mechanism is nucleophilic attack by OH^- ions as shown by the arrows in Fig. 1–2 and the first product formed is cyclic 2′,3′-phosphate (so called because the 2 and 3 positions of the ribose ring are esterified). The cyclic phosphate is only an intermediate in this reaction and is attacked by further OH^- ions cleaving the C—O—P bonds with equal facility and hence producing a mixture of 2′- and 3′-phosphates. From this it can be seen that, under conditions of mild alkaline hydrolysis, the breakdown of RNA is brought about in several steps in a partially controlled manner.

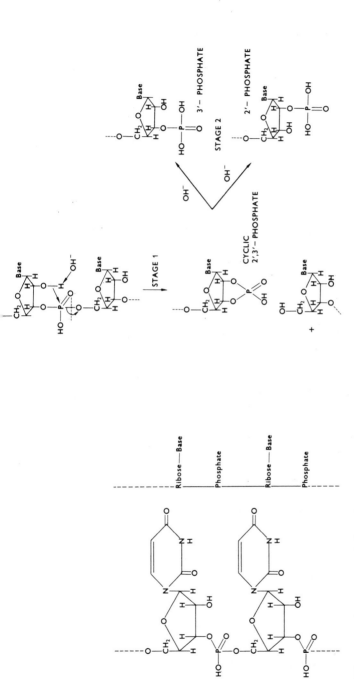

Fig. 1-1 A segment of the ribonucleic acid molecule (RNA) showing the nature of the bonding and also a summary of the relative positions of the component moieties. In this example only one type of base, uracil, is shown although in the normal RNA molecule three other bases, adenine, guanine and cytosine also occur.

Fig. 1-2 The mechanism of mild alkaline hydrolysis of RNA.

1.3   Ribonuclease and its mechanism of action

It is a fundamental property of living systems that they are in a state of dynamic equilibrium, i.e. that the molecules which make up the chemical skeleton of the cell are constantly being synthesized and degraded. Ribonucleic acid is no exception and there are present in all cells enzymes, ribonucleases, which hydrolyse this molecule. The chemical structure and physical properties of these enzymes have been investigated extensively and from these studies have emerged various theories of the mechanisms of the reactions catalysed.

The ribonuclease of bovine pancreas has been shown to hydrolyse the RNA in two stages in a manner analogous to the action of alkali. The first stage is the production of the cyclic phosphate and this is followed by the breakdown to the 3′-phosphate. Lest it be thought that this can be simply explained by the presence of OH^- ions it is pertinent to note that the enzyme acts in buffer of approximately neutral pH where the OH^- ion concentration is extremely small.

A possible mode of action of this enzyme is shown in Fig. 1–3. There are at least two ionizable groups in the enzyme which are involved in the catalytic activity and which form part of what is termed the 'active centre' of the enzyme. In the first stage of the reaction, group A serves as a proton acceptor paralleling the basic character of the OH^- ion while group B acts as a proton donor, i.e. it behaves as an acid. In the second stage of the reaction, the group A is now a proton donor while B is proton acceptor. After the complete series of reactions the enzyme is in its original state and ready to perform the catalysis of the hydrolysis of another P—O bond.

In a way the enzyme has brought together both acid and base hydrolysis and is consequently more effective than either separately. Of course it is impossible to devise a conventional non-biological system containing both acid and base since neutralization would occur. The presence of these two ionizable groups at the active centre make the enzyme especially suited for its hydrolytic role.

1.4   Structure and biological function

This example of the mechanism of action of the pancreatic ribonuclease shows that there is a logical and strictly chemical explanation of the action of an enzyme. This has been found to hold true for all the enzymes whose mode of action has been studied in detail and there is no reason to believe that it is not universally applicable.

It is implicit in the reaction scheme given in Fig. 1–3 that the groups A and B are so arranged in space that they are in close proximity to the bond in RNA which is being hydrolysed. The groups, being part of a large molecule, are not free to diffuse through the solution of RNA as could OH^- ions and are thus dependent on being correctly spatially aligned with

1 FORMATION OF CYCLIC PHOSPHATES

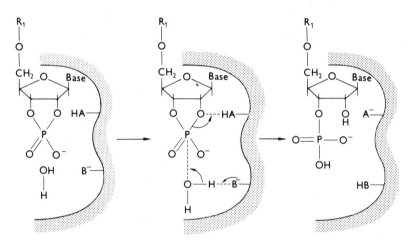

2 HYDROLYSIS

Fig. 1-3 A possible mechanism for the action of pancreatic ribonuclease on RNA. The enzyme with its two catalytic groups A and B is shown by shading. Only one repeating sequence of the RNA molecule is shown and R_1 and R_2 represent the remainder of the molecule.

respect to each other and also with respect to the RNA molecule which must be bound to the enzyme in some way. The activity of the enzyme is dependent on the three-dimensional structure of the molecule and before the action of an enzyme can be fully appreciated it is necessary to consider the various chemical bonds and physical forces which together give the enzyme its structure and stability.

Protein Structure 2

2.1 Enzymes are proteins

As soon as detailed analysis of the chemical structure of enzymes was undertaken, it was recognized that all enzymes were proteins. Today, even though over one thousand enzymes have been discovered and many of them fully characterized, there is no evidence to throw doubt on the validity of the identity of enzymes as proteins. In some cases there are additional features in the enzyme molecule such as metallic ions or carbohydrates and other small organic molecules but the essential catalytic function of the enzyme is always dependent on the protein moiety, although these other non-proteinaceous materials may contribute, and in some cases be essential, to the catalytic activity.

The converse is not true for some proteins are apparently devoid of catalytic activity. Collagen, the major protein of bone, skin and tendon is an example of such a protein. Collagen seems to play a structural role in the body and as such might not be expected to take part in the degradative and synthetic processes, which are collectively termed metabolism. Another example of a structural rather than a metabolically active protein is keratin which forms part of the exterior surfaces of the body such as hair, nails, horns and the outer layer of skin—the epidermis. Serving a different function are the proteins called γ-globulins. These are the antibodies which provide part of the body's defence mechanism against disease and all invasions by foreign matter. γ-Globulins have not been shown to possess any catalytic activity.

To understand fully the nature of enzymes it is necessary to consider the detailed structure of proteins. Proteins are polymers of amino acids, of which there are twenty which commonly occur in proteins. From this relatively small number of basic building units it must be possible to make at least a thousand different proteins and so obviously the nature and quantity of each amino acid, together with its arrangement within the protein polymer will be crucial to the structure of the enzyme. We must first consider the nature and variety of amino acids.

2.2 The nature and diversity of amino acids

The amino acids possess both amino and carboxyl groups and therefore exhibit both basic and acidic properties. They may be represented by the general formula:

$$\begin{array}{c} R \\ | \\ H_2N\!-\!\!\!-\!\!C\!-\!\!H \\ | \\ COOH \end{array}$$

where R is different for each of the different amino acids. Except in the case of the simplest amino acid, glycine, where R is hydrogen, the central carbon atom is bonded to four different groups thus giving asymmetry to the molecule and hence the ability to rotate the plane of polarized light, i.e. optical rotation. Proteins also show optical activity, both as a result of their constituent amino acids and also as a result of the further asymmetry which is produced in the molecule of polymer by the linkage of the amino acids. This in part accounts for the extreme selectivity in the reactions which they will catalyse and also the precision of the products formed.

Later in this chapter we shall see that the amino and carboxyl groups are involved in the polymerization of amino acids to form protein. Thus it is the nature of R which contributes most to the catalytic activity of proteins and the diversity of R which allows the diversity seen in the variety of reactions catalysed. It is convenient to classify the various R groups in seven broad categories (Table 1). The seventh category, the

Table 1   Classification of amino acids by the nature of their side chains (R)

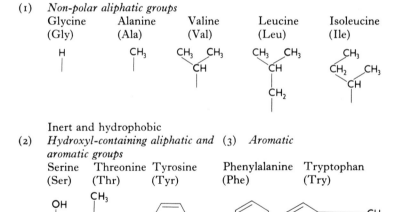

(1)   *Non-polar aliphatic groups*

| Glycine (Gly) | Alanine (Ala) | Valine (Val) | Leucine (Leu) | Isoleucine (Ile) |

Inert and hydrophobic

(2)   *Hydroxyl-containing aliphatic and aromatic groups*   (3)   *Aromatic*

| Serine (Ser) | Threonine (Thr) | Tyrosine (Tyr) | Phenylalanine (Phe) | Tryptophan (Try) |

Aliphatic are neutral but tyrosine is weakly acidic. Less hydrophobic than the non-polar aliphatic amino acids.

Tyrosine is also aromatic but is conveniently classed with the other hydroxyl-containing amino acids. Hydrophobic.

(4) *Acidic*

Aspartic acid (Asp)	Glutamic acid (Glu)	Asparagine (Asn)	Glutamine (Gln)

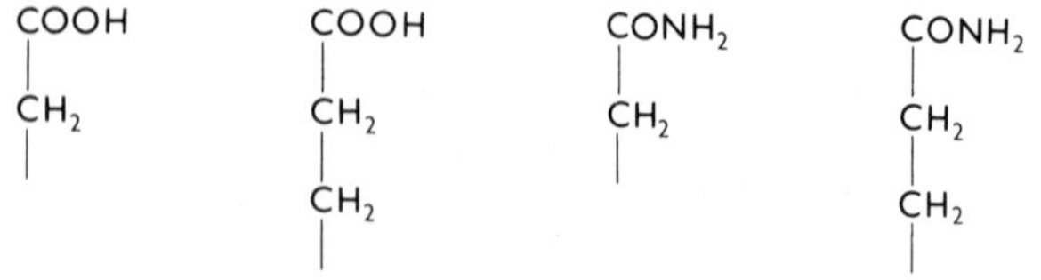

$$
\begin{array}{cccc}
\text{COOH} & \text{COOH} & \text{CONH}_2 & \text{CONH}_2 \\
| & | & | & | \\
\text{CH}_2 & \text{CH}_2 & \text{CH}_2 & \text{CH}_2 \\
& | & & | \\
& \text{CH}_2 & & \text{CH}_2 \\
& | & & |
\end{array}
$$

Asparagine and glutamine are the amides derived from aspartic acid and glutamic acid respectively.
Hydrophilic.

(5) *Basic Groups*

Lysine Arginine Histidine
(Lys)   (Arg)   (His)

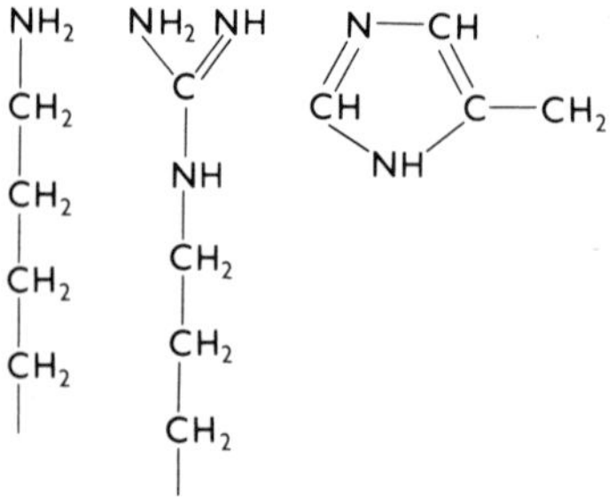

(6) *Sulphur containing groups*

Cysteine Cystine Methionine
(Cys)   (Cys-Cys) (Met)

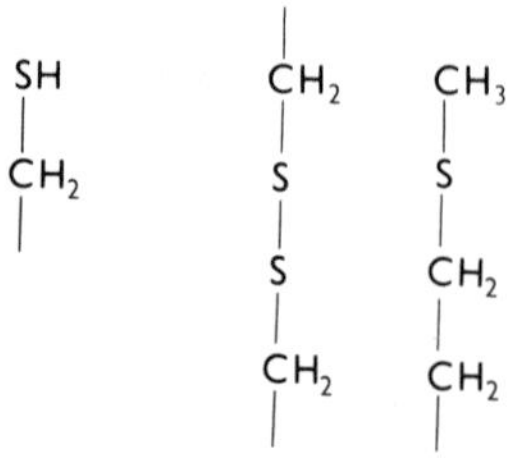

Cysteine is weakly hydrophilic while methionine is hydrophobic. Cystine is two cysteine residues linked by a S—S bond

Hydrophilic

(7) *Imino groups*

Proline
(Pro)

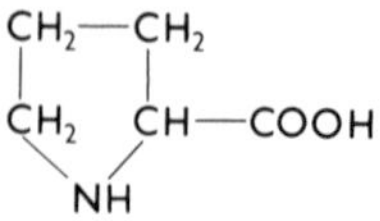

Hydroxyproline
(Hyp)

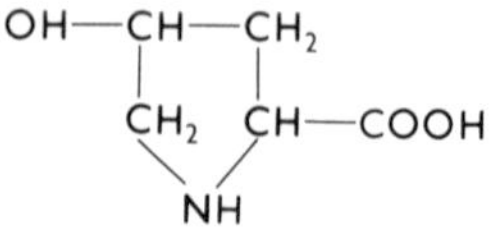

$$
\begin{array}{cc}
\text{CH}_2\text{—CH}_2 & \text{OH—CH—CH}_2 \\
| \qquad | & | \qquad | \\
\text{CH}_2 \quad \text{CH—COOH} & \text{CH}_2 \quad \text{CH—COOH} \\
\diagdown \; \diagup & \diagdown \; \diagup \\
\text{NH} & \text{NH}
\end{array}
$$

In this case the full amino acid structure is shown.

amino acids, are not true amino acids since in this case the amino group is not free but forms part of a ring structure. This difference leads to important modifications of protein structure as we shall see later. The important physical properties of the group are summarized at the end of each group. Of particular note is the interaction of the group with water.

It is possible to divide groups roughly into two categories:

(1) *Hydrophilic—literally 'water loving'*   Such groups confer the property of easy solubility in water. Acidic and basic groups are very hydrophilic.

(2) *Hydrophobic or 'water hating'*   Such groups are not easily soluble in water and other polar solvents such as alcohol but are readily soluble in ether, benzene and similar non-polar solvents due to the hydrocarbon nature of the group.

2.3   Primary structure—the peptide bond and the sequence of amino acids

Elimination of water between the carboxyl group of one amino acid molecule and the amino group of another results in the formation of a linkage similar to the type found in amides and called in this case the peptide bond:

$$\underset{R_1}{H_2N.CH.COOH} + \underset{R_2}{H_2N.CH.COOH} \longrightarrow \underset{R_1 \quad R_2}{H_2N.CH.CO.NH.CH.COOH} + H_2O$$

where R_1 and R_2 represent the same or different substituent groups. The molecule formed in the above reaction is termed a dipeptide. Because the amino acid is bifunctional (i.e. it contains both an amino and a carboxyl group) this process may continue leading to the formation of a tripeptide, tetrapeptide, etc. After the condensation of a large number of amino acids the product is known as a polypeptide. The precise distinction between a polypeptide and a protein is not clear. Most proteins have at least 100 amino acids linked together (commonly expressed as amino acid residues) and in addition have a definite higher order structure. Polypeptides, on the other hand, is a term usually confined to molecules containing of the order of twenty amino acid residues and lacking a higher order of structure. Each polymeric molecule will only have one free α-amino and one free α-carboxyl group. Thus the properties of these two groups will play little part in the properties of the protein molecule, which will depend very largely on the properties of the R group. As can be seen from the table some amino acids contain additional amino and carboxyl groups and these will of course be free and contribute to the protein's properties.

Although the equation above adequately represents the chemistry of the formation of the peptide bond it gives no idea of the three dimensional orientation of this bond. Figure 2–1 shows that the two α-C atoms and the intervening atoms of C,O,N,H all lie in the same plane since the angles between the bonds joining these atoms add up to 360°. It is only around the α-C atom which has four different groups attached to it, that free rotation can take place. This is partly explained by the partial double bond character of the C–N bond in the peptide, as shown by the length of the bond which is less than that of C—N in an amino acid. It can also be seen

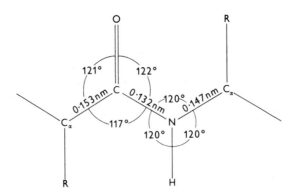

Fig. 2.1 The geometry of the peptide bond.

that the R groups are arranged in a trans configuration along the peptide backbone.

The order in which the amino acids are joined together is called the primary sequence of a protein and this together with the structure of the peptide bond make up what is termed the primary structure of the protein. The sequence of amino acids in many proteins has now been determined. The full structure of the hormone corticotrophin as isolated from sheep is:

Ser.Tyr.Ser.Met.Glu.His.Phe.Arg.Try.Gly.Lys.Pro.Val.Gly.Lys.Lys.
Arg.Arg.Pro.Val.Lys.Val.Try.Pro.Ala.Gly.Glu.Asp.Asp.Glu.Ala.Ser.Glu.
Ala.Phe.Pro.Leu.Glu.Phe.

The abbreviations of the amino acids are usually used in writing the sequence of a protein or polypeptide and it is convention that the sequence should be written from the N-terminal end, i.e. from the end that has the free α-NH₂ group.

What would happen if the same 39 amino acids were joined together in a different order? In general the molecule would lose its biological activity. Thus the nature and sequence of amino acids are always specific for a particular protein. When one considers proteins serving identical functions for different species then some slight variations in the nature and sequence of the proteins is seen. For the corticotrophin the residues 1–24 and 33–39 appear to be invariable for all species studied. However, minor modifications are apparent in the residues 25–32 when the primary sequences of pig, human, ox and sheep corticotrophins are compared. Deviations from

	25	26	27	28	29	30	31	32	33
Pig	Asp	Gly	Ala	Glu	Asp	Gln	Leu	Ala	Glu
Man	Asp	Ala	Gly	Glu	Asp	Gln	Ser	Ala	Glu
Ox	Asp	Gly	Glu	Ala	Glu	Asp	Ser	Ala	Gln
Sheep	Ala	Gly	Glu	Asp	Asp	Glu	Ala	Ser	Gln

the sequence of pig corticotrophin are underlined. Very often the alterations are of a minor nature and amino acids with similar properties are involved. Thus alanine has replaced glycine to which it is very similar and glutamic acid has replaced the other acidic amino, acid, aspartic acid.

2.4   The active centre concept

It is now well established that only a few specific amino acid residues in a protein molecule are involved directly in the functioning of that protein. In the case of enzymes, these specific residues are responsible for the catalytic activity of the molecule. The other amino acid residues help to form and stabilize the three-dimensional structure of the molecule as we shall see later. In the case of corticotrophin, residues between 1 and 24 and/or between 33 and 39 are probably responsible for the hormonal function while the variable residues between 25 and 32 help to direct the general shape of the molecule. They may also be involved in maintaining the individuality of the species whereby each species is able to recognize protein molecules of its own species.

Figure 2–2 shows the primary sequence of the protein, ribonuclease, which was introduced in the first chapter. Ribonuclease contains 124 amino acids and it has been shown that the histidines at 12 and 119 (written His 12 and 119) together with Lys 7 and 41 are involved in the actual reaction catalysed. Now these residues are widely separated as the structure is written in this figure. Yet in the active ribonuclease molecule they must be fairly close together since the attack is on a single bond in the RNA molecule. This group of amino acid residues constituting the directly functional part of the molecule is termed the active centre. It is obvious that, in order to bring together residues which are far removed in a linear sequence, there must be a folding of the protein molecule.

2.5   Disulphide bridges

The amino acid, cysteine, contains a thiol group and such groups are readily oxidized to disulphides in which the two sulphur atoms are joined together. Such disulphide bridges are readily formed between cysteine residues in proteins. They may serve to link two protein molecules together or, if the protein is large and contains at least two cysteine residues, they may serve to link two parts of the same molecule together, thus creating a permanent fold in the structure. The structure of the disulphide linkage is

$$\begin{array}{ccccccc} | & & & & & & | \\ NH & & & & & & NH \\ | & & & & & & | \\ CH & -CH_2 & -S & -S & -CH_2 & -CH \\ | & & & & & & | \\ CO & & & & & & CO \\ | & & & & & & | \end{array}$$

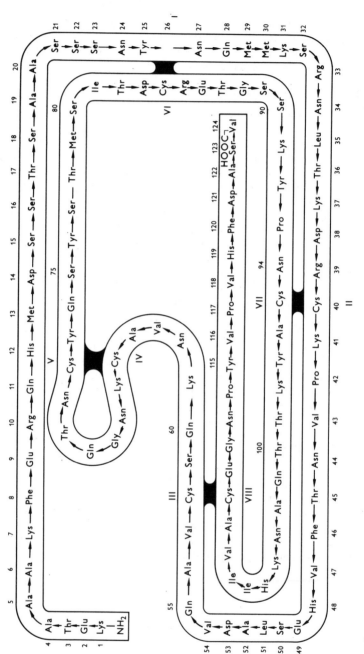

Fig. 2-2 The primary sequence of bovine pancreatic ribonuclease A. His 12 and 119 and Lys 7 and 41 form part of the active centre. (Reproduced with permission from SMYTH, STEIN and MOORE, 1963. *J. Biol. Chem.* **238**, 228.)

From Fig. 2–2 it can be seen that ribonuclease contains four such disulphide linkages involving eight cysteine residues. These disulphide bonds limit the free rotation of the residues in ribonuclease and lead to folding of the chain. However, this does not adequately explain how the various groups in the active centre are brought together. An adequate explanation is only possible when the secondary and tertiary structure of the enzyme are considered.

2.6   Secondary structure: α-helix, β-pleated sheets and hydrogen bonding

Secondary structure is the term used to describe the folding of the otherwise flexible peptide chains resulting from the formation of hydrogen bonds between the carbonyl oxygen and amide nitrogen atoms of the polypeptide backbone. It must be emphasized that it is a specific type of structure and only involves hydrogen bonding between atoms in the backbone. Hydrogen bonding between atoms in the sidechain amino acid residues do not contribute to the secondary structure but as we shall see later they are very important in determining tertiary structure.

Hydrogen bonds are very weak bonds having an energy of approximately 16.74 kJ/mole compared to the energy of a covalent C—H bond which is 364 kJ/mole. However, if there are sufficient hydrogen bonds in a molecule these impart a reasonable stability to the structure. The hydrogen bond and its effect on the physical properties of a molecule are best considered in water. The electrons in the covalent O—H bond in water are not equally shared between the atoms but are drawn towards the O atom thus giving the bond a polar character. This may be represented by a partial negative charge on the O atom and a partial positive charge on the hydrogen atom. These partial charges cause electrostatic interaction between different molecules of water as shown in Fig. 2–3. Since the oxygen has six electrons in its outer shell only two of which are used in the water molecule for bonding, the other electrons have a tendency to be given to the hydrogen atom which is electron deficient. Although this does not lead to full bond formation, it helps to explain the strength of the hydrogen bond. The overall effect of this is that water cannot be considered as an H_2O molecule but must be thought of as a loose association of such molecules. From the position of oxygen in the periodic table it would be expected that water would be a gas at room temperature (cf. H_2S which is a gas). It is the association of the molecules by means of hydrogen bonding that gives water the property of being liquid at these temperatures.

Not only the H atom in O—H bonds is able to form hydrogen bonds. In a situation where hydrogen is bonded to a dissimilar atom and that atom by electron attraction creates a polarity in the bond, then hydrogen bonding can occur. In proteins the most important hydrogen bonds are: —N—H . . . O— and N—H . . . N— where the dotted lines represent hydrogen bonds. It has been shown that the bond is most stable when the N,H and O or N atoms lie in a straight line.

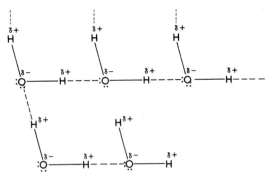

Fig. 2-3 The structure of liquid water. Hydrogen bonding is shown by dotted lines.

Secondary structure which arises because of hydrogen bonding between various parts of the polypeptide backbone falls into two broad categories: the α-helix and the β-pleated sheet. The α-helix was first devised from theoretical considerations by Pauling and Corey. They argued that the most stable folding of a protein molecule must fulfil at least four basic criteria:

(1) the maximum number of hydrogen bonds should be formed between CO and NH groups in the peptide backbone.

(2) the peptide bond should be planar as shown in Fig. 2-1.

(3) the O,H and N atoms of the hydrogen bond shouldl ie in a straight line.

(4) there should be a constant spatial progression of atoms as you move along the structure, i.e. C,N and C atoms should repeat at regular intervals in a linear direction.

Figure 2-4 shows the helical structure which they built up to fulfil such criteria. It has been called the α-helix and has several important features. There are 3.6 residues per turn leading to a repeating pattern every 5 turns. The angle of the helix is 26° and the pitch is 0.54 nm. This structure has subsequently been confirmed by the X-ray crystallographic analysis of protein crystals—a technique to be discussed later. In order to emphasize the planarity of the peptide bond a diagrammatic representation of the α-helix is shown in Fig. 2-5 where for clarity the individual atoms are omitted. The imino acids, proline and hydroxyproline cannot be accommodated in the α-helical structure and hence this ordered structure breaks down in regions of the protein chain where these imino acids occur.

Figure 2-6 shows the hydrogen bonding in a β-pleated sheet. Two forms of this structure are possible. Segments of the polypeptide may be brought together so that the chains are parallel and running in the same direction as shown by the order of the α-C atom, the carbonyl and NH groups. Such a structure is termed the parallel β-pleated sheet. Alternatively, the chains though parallel may run in opposite directions. This type of secondary structure is known as the anti-parallel β-pleated sheet. Sometimes the pleated sheet structure is used to link adjacent chains of different polypeptide molecules. This is often the case for fibrous structures like the protein, fibroin, found in silk and the protein, keratin, found in hair. Figure 2.7 again emphasizes the planar nature of the peptide bond and shows how the pleated structure can be built up into a large three-dimensional array.

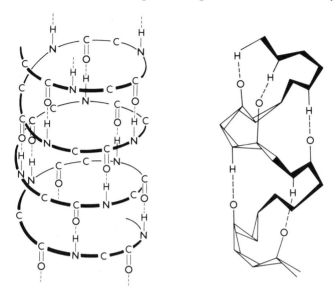

Fig. 2–4 (*left*) The α-helix showing the stabilization of the structure by intra-chain hydrogen bonding. (Reproduced with permission from DEARDEN, 1968. *New Scientist*, **37**, 629.)

Fig. 2–5 (*right*) The α-helix showing the planarity of the peptide bond. The individual atoms have been omitted for clarity. (Reproduced with permission from BARKER, 1968. *Understanding the chemistry of the cell*, Edward Arnold, London.)

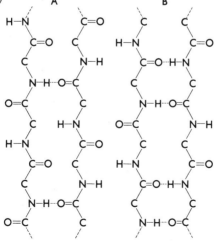

Fig. 2–6 A. Parallel β-pleated sheet. B. Anti-parallel β-pleated sheet. For clarity the R substituent on Cα is omitted.

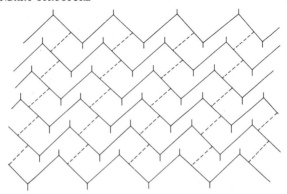

Fig. 2–7   Three dimensional network of β-pleated sheet showing the planarity of the peptide bond which is perpendicular to the plane of the paper.

Although most proteins show one or both of these structures, by no means all of the peptide backbone is arranged in this orderly fashion. Some proteins may contain as much as 80% ordered structure while for others the figure may be as low as 20%. Very often small sections of helical coiling are seen separated by large amounts of apparently random folding of the peptide backbone. In fact this apparent randomness is an ordered structure imposed by the tertiary structure of the molecule. Among the forces that govern the formation of tertiary structure are electrostatic forces. We shall now consider the origin of these electrostatic forces between the side-chains of the amino acid residues.

2.7   The ionization of amino acids and proteins

The tendency of an acid to ionize giving rise to a proton is dependent on the proton concentration of the solution. Thus if one considers the ionization of a weak acid such as acetic acid then equilibrium is established as shown in the following equation:

$$CH_3\ COOH\ +\ H_2O\ \rightleftharpoons\ CH_3COO_3^-\ {}^+H_3O^+$$

At low pH values (e.g. a solution of acetic acid in a concentrated mineral acid) the solvated proton concentration is very high and negligible ionization occurs. On the other hand at high pH values (e.g. acetic acid dissolved in some basic solution) the solvated proton concentration is very low and the ionization is virtually complete. At some intermediate pH value the position of the equilibrium is such that the concentration of the ionized and unionized acetic acid are the same. This pH is termed the pK of the ionizing species. Thus the pK of acetic acid is 4.7.

The carboxyl group is by no means the only ionizing group that is found in amino acids and proteins. At low pH values the amino group is able to accept a proton, a process represented by the following equation:

$$R.NH_3^+\ +\ H_2O\ \rightleftharpoons\ R.NH_2\ +\ H_3O^+$$

In the case of the simple aliphatic amine, methylamine, this equilibrium is such that there are equal concentrations of the ionized and unionized species at pH 10.7. Thus the pK of methylamine is 10.7.

Table 2 Ionizable groups in amino acids and proteins

Group	Equilibrium	Likely range of pK
α-carboxyl	$-COOH \rightleftharpoons COO^- + H_3O^+$	3.0–3.5
β & γ-carboxyl (Aspartic and glutamic acids)	$-COOH \rightleftharpoons COO^- + H_3O^+$	3.0–4.7
Imidazole (Histidine)	$NH_2^+ \rightleftharpoons NH + H_3O^+$	5.6–7.0
α-amino	$NH_3^+ \rightleftharpoons NH_2 + H_3O^+$	7.6–8.4
Sulphydryl	$SH \rightleftharpoons S^- + H_3O^+$	8.3–8.6
Phenolic OH	$OH \rightleftharpoons O^- + H_3O^+$	9.8–10.4
ε-amino (Lysine)	$NH_3^+ \rightleftharpoons NH_2 + H_3O^+$	9.4–10.6
Guanidino (Arginine)	$NH_2^+ \rightleftharpoons NH + H_3O^+$	11.6–12.6

Table 2 summarizes the ionizations of important groups in amino acids and proteins and also gives a range within which the pK value for the ionizing group is likely to lie. The pK of any particular group is dependent to some extent on the rest of the molecule. Although the pK of acetic acid is 4.7, the pK of the carboxyl group in glycine is 2.34. Similarly the ionization of the amino group has a pK value of 9.6. The same situation exists for proteins where the micro-environment of the ionizing group can lead to wide fluctuations in the value of pK. Of course in proteins, most of the α-carboxyl and α-amino groups are involved in peptide bond formation and the other ionizable groups assume a greater importance.

Table 3 The ionic forms of lysine

Ionic species	NH_3^+ $(CH_2)_4$ $^+H_3N.CH.COOH$	NH_3^+ $(CH_2)_4$ $^+H_3N.CH.COO^-$	NH_3^+ $(CH_2)_4$ $H_2N.CH.COO^-$	NH_2 $(CH_2)_4$ $H_2N.CH.COO^-$
Charge	+2	+1	0	−1
pH	Less than 2.5	4.5 to 7	9.0	Greater than 10

The implication of these ionizations in the protein under normal conditions is best seen by considering the ionization of lysine at various pH values. Table 3 summarizes the various ionic forms of lysine to be found over a range of pH varying from extreme acid to extreme alkaline conditions. It can be seen that the charge varies with pH and that the molecule is charged over the physiological range of pH, i.e. from 5 to 9. This is particularly true when the lysine is part of a protein structure and the α-amino and α-carboxyl groups are not available for ionization. Throughout the whole of the physiological range the ϵ-amino group will carry a positive charge. In the same way that carboxyl group in aspartic and glutamic acids, which is not involved in peptide bond formation, will have a negative charge throughout this range.

It is found that most proteins have only a small net charge at physiological pH values. Indeed the isoelectric point of a protein, i.e. the pH at which the ionic charge on the protein has net value of zero, is mostly found in the region 4.5 to 8.5. However, it must be emphasized that, locally in the protein molecule, the charge may be considerable due to the presence of several charged residues. Much of the tertiary structure of the protein depends for its integrity on the maintenance of these charges and similarly the catalytic activity of an enzyme is mostly dependent on the state of ionization of the side chains of amino acid residues at the active centre.

2.8   Tertiary structure

Over and above the limitations placed on the size and shape of the protein molecule by the nature of its primary and secondary structure, each protein molecule in solution has a unique and characteristic shape which is called its tertiary structure. This shape is the result of many intermolecular forces and represents the coiling and folding of the protein molecule so that the most stable structure is formed. Four main interactions between the side chains of amino acid residues can be considered:

(1)   ELECTROSTATIC INTERACTIONS BETWEEN CHARGED SPECIES   The well established phenomenon of attraction between unlike charges and repulsion between like charges ensures that certain regions of the protein molecule are brought together by electrostatic interaction. Amino acids that may be involved in such interactions include histidine, lysine, arginine, aspartic and glutamic acids, cysteine and tyrosine.

(2)   HYDROGEN BONDING   Serine, threonine and tyrosine with their OH groups are particularly important in the formation of hydrogen bonds and side chain carboxyl groups are also involved.

(3)   HYDROPHOBIC AND HYDROPHILIC INTERACTIONS   As described previously the side chains of amino acid residues may be broadly classified according to their interaction with water. Of particular importance in the maintenance of tertiary structure are the hydrophobic groups. These groups tend to interact to create a non-aqueous environment in the centre of the protein. In this way water can be excluded and the hydrophobicity of the groups satisfied by a protein in aqueous solution. Thus when the tertiary structure of a protein is examined in detail it is common to find the majority of non-polar hydrophobic groups in the interior of the structure and conversely the polar hydrophilic groups are often found on the exterior surface of the structure.

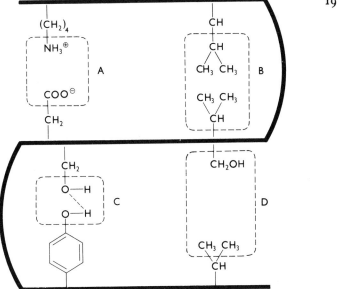

Fig. 2-8 Interactions involved in tertiary structure. A. Electrostatic. B. Hydrophobic. C. Hydrogen bonding. D. Van der Waals forces.

(4) VAN DER WAALS FORCES All atoms of all molecules interact to some extent with the atoms of other neighbouring molecules and one can define a distance of closeness of approach of individual atoms. This distance is dependent on the value of the nuclear charge and the nature of the electron distribution around that nucleus. The closeness of approach of the atoms of the protein chain as they coil and fold will be determined by such considerations. These interactions are summarized in Fig. 2-8. The tertiary structure of the protein represents the summation of all these effects such that the most stable structure is formed under any given set of conditions. This structure represents the state of lowest energy of the system.

There may be other very similar energy states of the system and a slight alteration in the conditions may allow the tertiary structure to alter. Thus the effect of heat is to disturb the tertiary structure causing a recoiling and refolding which may lead to a loss of enzymic activity.

Returning to ribonuclease, Fig. 2-9 shows the outline of the tertiary structure of this enzyme. It can be seen that the combined effect of the secondary and tertiary structure has brought together the various groups involved in the active centre, thus providing the necessary orientation for catalytic activity. This representation of the molecule emphasizes the folding of the backbone and does not give a true impression of the compactness of the structure. It does however emphasize the cleft in the molecule where the RNA molecule fits.

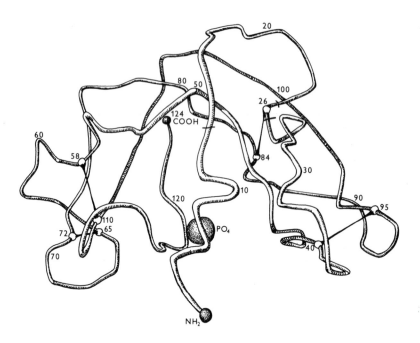

Fig. 2–9   The tertiary structure of ribonuclease. The histidines at 12 and 119 and the lysines at 7 and 41 have been brought together at the active centre. (Reproduced with permission from KARTHA, BELLO and HARKER, 1967. *Nature*, **213**, 864.)

2.9   Quaternary structure

In recent years it has been recognized that many proteins are not normally present as single folded polypeptide chains but that combinations of these single chains, called monomers, occur to give a complete structure with multiple monomer units. This complete description of the protein is called the quaternary structure. Various numbers of monomers have been found in functional proteins, e.g. dimers, trimers and tetramers occur regularly and there are reports of higher degrees of organization such as protein molecules containing twelve sub-units. In the more complex cases the quaternary structure may be made up from two different sub-units. It is common to find that when the protein is dissociated into monomeric chains the molecule has altered properties. For instance, it may respond to different substrates, it may change its optimum working conditions or in the most extreme case it may lose completely its catalytic activity. In general, it is possible to restore activity by allowing the recombination of monomers to reform the quaternary structure. In nature, it has been shown that there is an equilibrium set up between the monomer and the polymer and it is obvious that the factors which influence this equilibrium will in turn modify and control the catalytic activity. In chapter 5 we shall see examples of the control of metabolism by control of the sub-unit structure of the enzymes involved.

Having described the details of protein structure, we must now turn our attention specifically to enzymes and consider those properties and methods which are peculiar to these specialized protein molecules.

Nomenclature, Coenzymes and Methods in Enzymology

3.1 Nomenclature

When the first enzymes were discovered and partially characterized, the nomenclature was left to the agreement or sometimes disagreement of the individual investigators. Names current at this period reflected many facets of the study and properties of the enzymes. The name trypsin for the pancreatic proteolytic enzyme was derived from the Greek word meaning 'rubbing' and reflected the extraction of the enzyme from pancreas by grinding with various solvents. Other enzymes were named after their discoverers or reflected some obvious physical property of the enzyme e.g. 'old yellow enzyme'. In some cases an attempt was made to include in the name some indication of the nature of the reaction catalysed. One of the earliest discoveries was an enzyme from jack bean meal called urease. Although this name indicates that the substrate whose reaction is catalysed is urea, it gives no information on the nature of the products or of the type of reaction catalysed. It does however reflect the growing use of the convention of using the suffix '—ase' on the substrate to name the enzyme during the early part of the twentieth century.

With the discovery of many hundreds of enzymes, some of which were able to catalyse different reactions of the same substrate, it became imperative to reconsider the basis of nomenclature of enzymes. The International Union of Biochemistry set up a committee to investigate the problem and their findings were finally reported in 1964. It was decided that the name of the enzyme should reflect the nature of the substrate and the type of reaction catalysed, all information to be based on the formal chemical equation describing the reaction. To facilitate the nomenclature six broad types of reaction were defined and all known enzymes assigned to one of these groups. This system, although rather arbitrary and in some cases anomalous, nevertheless has proved valuable and is now in common usage. As some of the systematic names are rather clumsy, where no reasonable possibility of confusion can exist, the use of the previous trivial names is still allowed.

The six categories of enzymes defined were as follows:

(1) Oxidoreductases—catalysing all reactions where the reaction can be thought of as being of the oxidation-reduction type.

(2) Transferases—catalysing the transfer of a group from one substrate to another.

(3) Hydrolases—catalysing hydrolytic reactions.

(4) Lyases—catalysing additions to a double bond or removal of a group from a substrate without hydrolysis often leaving a compound containing a double bond.

(5) Isomerases—catalysing reactions where the net result is an intramolecular rearrangement.

(6) Ligases—catalysing the formation of bonds between two substrate molecules using energy derived from the cleavage of a pyrophosphate bond such as in adenosine triphosphate (ATP).

Before discussing detailed examples of the use of this classification, it is necessary to consider the role of some specific non-protein molecules in enzymic reactions. Such molecules include ATP and are called coenzymes.

3.2 Coenzymes

Many enzymes, covering a wide range of reactions, require the presence of small non-protein molecules, called prosthetic groups, for catalytic activity. These molecules may be firmly bound or they may be only loosely associated with the enzyme. Since they are closely concerned in the actual catalysis, they are also known as cofactors or coenzymes. Early workers tended to reserve the use of the term 'prosthetic group' for those molecules which were firmly bound to the enzyme and used coenzyme to denote a loose association. However, this distinction is purely theoretical and a full range of degrees of association with enzyme has now been recognized. The use of the term 'coenzyme' to cover all molecules, other than the enzyme protein, which are involved in the catalysis is more meaningful. The distinction between substrate and coenzyme is less well-defined. In many senses a coenzyme may be acting as a substrate. However, whereas a substrate is specific for a very limited number of reactions, a coenzyme may be involved in a large number of reactions of a similar type. In some cases the overall result of the reaction may be to leave the coenzyme in an unchanged form, but even if the coenzyme molecule is altered it is usual to find that reconversion takes place by subsequent reactions. Thus coenzymes may be defined as small non-protein molecules which serve a specific function in a variety of enzyme-catalysed reactions and are either unchanged at the end of the reaction or are regenerated by some subsequent process.

Consider the general field of oxidation-reduction reactions. The most common coenzyme concerned in these reactions is nicotinamide adenine dinucleotide (NAD). Nucleotides have the general structure:

Base—Sugar—Phosphate

and NAD being a dinucleotide has the structure:

Nicotinamide—Ribose—Phosphate—Phosphate—Ribose—Adenine

The detailed chemical structure of part of this coenzyme is shown in Fig. 3–1. The coenzyme exists in both a reduced and oxidized form and so the oxidized form can function in catalysis as a hydrogen acceptor. In equations the oxidized form is usually written as NAD^+ and thus we have:

$$NAD^+ + 2H \rightleftharpoons NADH + H^+$$

This hydrogen is normally derived from the substrate of the reaction and the coenzyme is seen to function as a hydrogen acceptor. The NADH formed then transfers the hydrogen to other acceptors in the electron

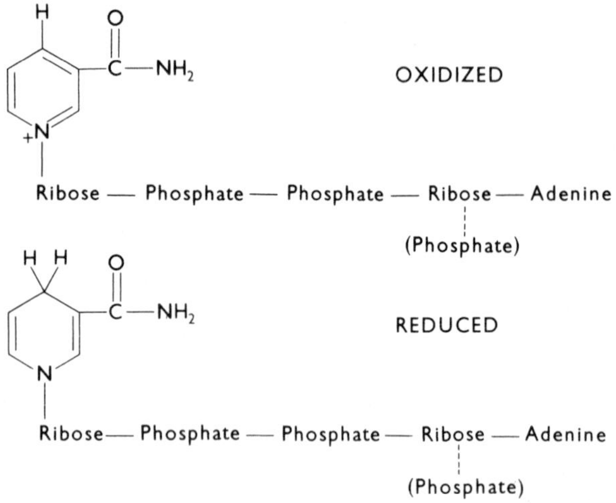

OXIDIZED

Ribose — Phosphate — Phosphate — Ribose — Adenine

(Phosphate)

REDUCED

Ribose— Phosphate — Phosphate — Ribose — Adenine

(Phosphate)

Fig. 3–1   The structures of oxidized and reduced NAD and NADP. The nicotinamide rings are shown in full to indicate the nature of reduction. The dotted lines indicate the position of the additional phosphate in NADP.

transport chain of the mitochondrion and NAD^+ is regenerated. The simplest example of a reaction involving NAD is the oxidation of alcohol to acetaldehyde by the enzyme alcohol: NAD oxidoreductase (trivial name —alcohol dehydrogenase):

$$CH_3CH_2OH + NAD^+ \rightleftharpoons CH_3CHO + NADH + H^+$$

The systematic name of the enzyme gives full information about the reaction. It specifies that the alcohol is undergoing an oxidation-reduction reaction and that the hydrogen acceptor is NAD.

Nicotinamide adenine dinucleotide acts as a hydrogen acceptor in a variety of similar reactions including the oxidation of alcohols to aldehydes, of aldehydes to acids and the desaturation of hydrocarbon chains. NAD is usually found as a hydrogen acceptor and the reverse use, i.e. NADH functioning as a hydrogen donor, is uncommon. The main reoxidation of NAD^+ is via the electron transport chain.

Closely related to NAD is the coenzyme nicotinamide adenine dinucleotide phosphate (NADP). It differs from NAD only in having an additional phosphate group on the ribose of the adenine part of the molecule (see Fig. 3-1). The function of NADP is rather different from that of NAD. It serves as a hydrogen acceptor in a limited series of reactions but acts as a hydrogen donor in a wide variety of reactions and may be thought of as the source of biological reducing power.

An entirely different type of function is exhibited by the coenzyme ATP. Adenosine triphosphate is again a nucleotide which possesses two additional phosphate groups. The arrangement of the parts is:

Adenine—Ribose—Phosphate—Phosphate—Phosphate

The detailed structure of ATP is shown in Fig. 3-2. ATP is the primary source of chemical energy in the cell. It is produced during the oxidation of carbohydrates, fats, amino acids and many other carbon compounds to carbon dioxide and water. These processes are analogous to combustion but, instead of the energy liberated being given out in the form of heat, the oxidation is coupled to the production of ATP, which is stored by the cell

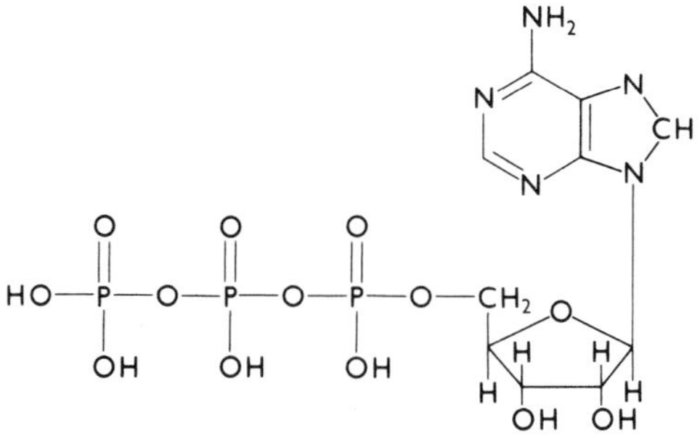

Fig. 3-2   The structure of ATP.

to provide a source of energy when needed. ATP is thus central to the energy metabolism of the cell. It occupies this special position by virtue of its energy rich properties. When the terminal phosphate group of ATP is hydrolysed to give adenosine diphosphate and inorganic phosphate, a large amount of free energy is released. This can be as much as 50kJ/mole depending on the intracellular conditions. The energy released by the hydrolysis of a simple carbohydrate phosphate ester, such as glucose-6-phosphate, to glucose and inorganic phosphate is considerably smaller, usually of the order of 16.7 kJ/mole. Because of this greater energy release on hydrolysis, ATP and several closely related compounds are termed 'energy rich compounds'. It must be emphasized that there is nothing especially magical about these compounds. It is purely the capacity for energy release on hydrolysis that warrants their separate consideration.

Occasionally the terminal phosphate bond is called a 'high energy phosphate bond' but the use of this term is misleading since it implies an unusual bonding between the oxygen and phosphorus atoms which is not true. In a similar way the energy released when ADP is hydrolysed to adenosine monophosphate (AMP) and inorganic phosphate is high. However when the last phosphate group is removed, i.e. when AMP is hydrolysed to adenosine and phosphate, the energy released is of the same order as that released on hydrolysis of a simple phosphate ester. The reasons for this phenomenon are extremely complex but rely essentially on the possibilities of resonance structures and the relative stabilities of the ATP, ADP and AMP.

One way in which ATP may function as a coenzyme is as a phosphate donor in an analogous manner to NADPH acting as a hydrogen donor. The conversion of glucose to glucose-6-phosphate is catalysed by the enzyme ATP: Glucose-6-phosphotransferase (trivial name glucokinase) as shown in the equation:

$$ATP + glucose \rightleftharpoons ADP + glucose\text{-}6\text{-}phosphate$$

Among the other coenzyme functions of ATP is to provide the energy necessary to form a covalent bond between two substrate molecules. The conversion of pyruvate and carbon dioxide to oxaloacetate can be represented by the formal equation:

$$CH_3CO.COOH + CO_2 + ATP + H_2O \rightleftharpoons HOOC.CH_2.CO.COOH$$
$$+ ADP + phosphate$$

The hydrolysis of ATP is coupled to the synthesis of a C—C bond in oxaloacetate. The enzymes catalysing such reactions are called ligases or synthesases and the specific enzyme in this case is pyruvate: carbon dioxide ligase (trivial name—pyruvate carboxylase).

A feature of the pyruvate carboxylase is that there is a second coenzyme involved in its catalytic activity. This other coenzyme is biotin which, unlike ATP, is bound firmly to the enzyme. Biotin which has a complex heterocyclic structure serves to bind carbon dioxide to the enzyme thus facilitating the combination of pyruvate and carbon dioxide. The ability to bind and transfer carbon dioxide is a common feature of the coenzyme function of biotin.

There are many other coenzymes each of them having an individual and specific use in metabolism as in the above examples. Some of the coenzymes are closely related to vitamins. Nicotinic acid is a direct precursor of the nicotinamide part of the NAD molecule. An adequate supply of this vitamin, nicotinic acid, in the diet is useful to man although in this case not absolutely essential, since the nicotinamide may also be derived from the amino acid tryptophan. In the case of some coenzymes the dietary intake of a vitamin precursor is absolutely essential to the normal functioning of metabolism.

3.3 Classification and numbering of enzymes

Table 4 shows the six categories of enzymes and indicates the way in which each class is subdivided. In addition a representative enzyme of each category is shown together with its number. This number has four com-

Table 4 Enzyme classification and numbering

---

1. Oxidoreductases.
    Sub-division: Nature of substrate, e.g. alcohol
    1.1.1.27   Lactate: NAD oxidoreductase
2. Transferases.
    Sub-division: group transferred, e.g. phosphate
    2.7.1.2   ATP: glucose-6-phosphotransferase
3. Hydrolases.
    Sub-division: bond hydrolysed, e.g. ester bonds
    3.1.3.9   Glucose-6-phosphate phosphohydrolase
4. Lysases.
    Sub-division: nature of substrate, e.g. amide
    4.3.1.1: Aspartate ammonia-lyase
5. Isomerases.
    Sub-division: type of isomerism, e.g. racemase
    5.1.1.1   Alanine racemase
6. Ligases.
    Sub-division: bond formed, e.g. C—C bond
    6.4.1.1   Pyruvate: carbon dioxide ligase

---

ponents. The first refers to the overall category of enzyme, the second and the third represent and sub- and sub-sub-class to which the enzyme has been assigned, and the fourth component represents the position of the enzyme within the sub-class. The detailed numbering of alcohol dehydrogenase is:

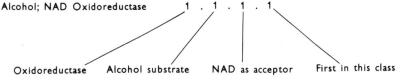

Alcohol; NAD Oxidoreductase   1 . 1 . 1 . 1

Oxidoreductase   Alcohol substrate   NAD as acceptor   First in this class

3.4 The measurement of enzyme activity

It is virtually impossible to measure the amount of any enzyme directly. In theory, if an enzyme contains some rare and specific feature such as a metallic ion, then measurement of that feature will give a measure of the amount of enzyme. However, this assumes the absence of any other component in the system containing the metal ion, and in all but the most pure enzyme preparations this asumption is unjustified. Hence methods of direct measurement of enzymes find no universal applicability.

In general, enzymes are measured by their catalytic activity, i.e. by measuring the rate of the reaction that they are catalysing and comparing it with the rate of the uncatalysed reaction. For most reactions the rate of the uncatalysed reaction is very low and for all practical purposes can be neglected. However, it is necessary to show that, under any given set of experimental conditions, the rate is low before making this approximation.

The methods of following reactions are as various as the number of enzyme-catalysed reactions known. However, there are certain experimental techniques which find wide applicability.

(1) *Spectrophotometry*   If, in the conversion of substrate to product, there is some specific and reasonably large change in the absorption properties of the system in the ultra-violet or visible region of the spectrum, then this change can be used to follow the reaction. This technique is particularly useful in following reactions where NAD is a coenzyme. Both NAD and NADH show a peak of maximum absorption at 260nm and there is only a small percentage difference between the absorption of equimolar solutions at this wavelength. However, NADH has another absorption peak at 340 nm and NAD has negligible absorption at this wavelength (Fig. 3–3). To study

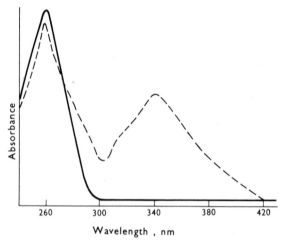

Fig. 3–3   Absorption spectra of NAD (full line) and NADH (dotted line).

alcohol dehydrogenase the absorption at 340nm of a suitable solution of alcohol and NAD is measured. At a given time a quantity of enzyme is added and the absorption of the system followed at 15s intervals over about 3 min. The rate of increase in absorption is a measure of NADH formed and hence a measure of the rate of catalysed reaction.

A unit of enzyme activity is defined as that amount of enzyme that will catalyse the conversion of one micromole of substrate to product in one minute under specified conditions. The specified conditions include pH, substrate concentration and temperature. Various standard temperatures have been suggested but in practice the temperature used is dependent on the biological origin of the enzyme under study. Thus enzymes from warm-blooded species are usually assayed (i.e. measured) at 37°C while enzymes from micro-organisms may be measured at lower temperatures, e.g. 25°C. In the alcohol dehydrogenase example, knowing the absorption of a molar solution of NADH, the amount of NADH formed in unit time can be found and hence the number of units of enzyme activity present.

In the study of hydrolytic enzymes, considerable use has been made of synthetic substrates derived from ortho- and para-nitrophenol. The enzyme arylsulphatase will hydrolyse p-nitrophenyl sulphate according to the equation:

$$O_2N\langle\bigcirc\rangle O.SO_3H + H_2O \rightleftharpoons O_2N\langle\bigcirc\rangle OH + H_2SO_4$$

The liberated p-nitrophenol at alkaline pH has a large absorption maximum at 400 nm while the organic sulphate has negligible absorption at this wavelength. If the enzyme will act under alkaline conditions or by terminating the reaction after a given period of time by the addition of strong alkali, the enzyme can be assayed by following the increased absorption at 400 nm. There are many similar substrates which can be used in this way.

(2) *pH measurement*   Many reactions, particularly hydrolyses, give rise to acidic products. Arylsulphatase gives sulphuric acid and so the reaction may be followed in unbuffered solutions with a sensitive pH meter. The inherent disadvantage of the technique used in this way is that the pH is constantly changing and this in turn is likely to affect the enzyme activity which is sensitive to pH change. To overcome this an apparatus called the pH-stat has been developed which monitors the pH and adds acid or alkali as appropriate to keep the pH constant, recording with time the amount of addition.

(3) *Manometric methods*   This technique is often used to study respiration, i.e. the uptake of oxygen and the simultaneous production of carbon dioxide by cell homogenates. The change in pressure of a system at constant volume is measured and by suitable refinements the oxygen uptake and carbon dioxide evolution can be calculated. This direct approach is of limited applicability to the study of individual enzymes because in very few cases is there any gas uptake or evolution. The technique may be extended to study reactions where acid is produced by including

CSFE

bicarbonate in the system and measuring the carbon dioxide evolved by the action of the acid.

In recent years this technique has been partly superseded by the development of the oxygen electrode which measures the oxygen tension of a solution in a manner analogous to the measurement of pH by the hydrogen electrode.

(4) *General chemical methods*   In some cases none of these specialized methods of enzyme measurement is applicable. In these cases it is necessary to use some general chemical method for following the disappearance of substrate or the appearance of product. For example, those enzymes which catalyse the hydrolysis of phosphate esters may be measured by stopping the catalysis after a known period of time, by altering the pH drastically or heat inactivating the enzyme, and then determining the phosphate liberated by standard phosphate analysis methods, such as forming the phosphomolybdate complex and reducing. Similarly, enzymes which catalyse protein hydrolysis may be measured by estimating the increase in free amino groups. It is usual to use colorimetric methods of chemical analysis in such cases because of the rapidity, convenience and sensitivity of such methods. In fact the spectrophotometer has become the enzymologist's most valuable tool.

3.5   Primary sequence determination

In order to understand fully the nature of any enzymically-catalysed reaction, it is necessary to elucidate completely the structure of the enzyme catalyst. The first stage of this process is the determination of the primary sequence of the molecule.

A prerequisite to the determination of primary sequence is the analysis of amino acids residues in the molecule. A pure protein is hydrolysed by treatment with 6M hydrochloric acid in an atmosphere of nitrogen in a sealed tube at 105°C for 24 h (or alternatively by autoclaving at 15 lb/sq. in. for four hours). The resultant solution of amino acids is evaporated to dryness under reduced pressure and then redissolved in a buffer of low pH. This solution is then allowed to flow through a column of ion-exchange resin. These ion-exchange resins are made from polymers, such as polystyrene, which have been substituted with various ionizable groups. The material is supplied in the form of beads which have been sieved to provide a narrow range of size which ensures an adequate and reproducible flow rate through a column of the resin. In addition during the manufacture of the resin the degree of crosslinking is carefully controlled. Those resins which have a high degree of crosslinking make a very tight-knit meshwork which only permits the passage of small molecules. Such a resin would be suitable for the separation of amino acids and like molecules but would be useless for proteins which would be trapped in the meshwork. All the amino acids have a net positive charge at low pH and are firmly bound to the negatively charged resin as they percolate through the column. For the separation of amino acids polystyrene substituted with sulphonic acid residues is commonly used. The sulphonic acid is a very strong acid and may be considered to be fully ionized.

$$Resin — SO_3H + H_2O \rightleftharpoons Resin — SO_3^- + H_3O^+$$

In acid solution the amino acid has the form: $^+H_3N.CHR.COOH$ and this sets up an equilibrium:

$$Resin—SO_3^- + {}^+H_3N.CHR.COOH \rightleftharpoons Resin—SO_3^-.{}^+H_3N.CHR.COOH$$

the equilibrium being very far to the right hand side. Thus the amino acids are electrostatically bound to the column. As the pH of eluting buffer is increased, the positive charge on the amino acid decreases until a point is reached where the binding of the amino acid to the resin is so weak that it is washed off the column. This point is reached at a different point for each amino acid and is dependent on the nature of other ionizable groups on the molecule and, even in the case of the non-polar amino acids, on the nature of the sidechain. Thus by absorbing the amino acids at a low pH and then slowly increasing the pH and ionic strength of the eluting medium, a clear cut reproducible separation of all the amino acids can be achieved. In modern practice the process is fully automated. The amino acid mixture is applied automatically to a column containing at least two types of resin for ease of separation. A gradient of pH and ionic strength is produced in the eluting buffer and the emergence of the amino acids from the bottom of the column is monitored by colorimetric determination which in turn is recorded on a chart. In the most elaborate instruments the colorimetric data are fed to a desk-top computer which automatically calculates the amount of the individual amino acids and the percentage of that amino acid in the original protein or more usually the residues/1000 residues.

Having established the amino acid composition of the enzyme, the next stage is to determine the amino acids at each end of the protein chain, i.e. the N- and C-terminal amino acid residues. The N-terminal amino acid is most conveniently determined by the use of 1-fluoro-2, 4-dinitrobenzene (FDNB) which reacts with amino groups to give a yellow dinitrophenylated amino acid (DNP-amino acid)

$$O_2N\langle\bigcirc\rangle F + H_2N.R = HF + O_2N\langle\bigcirc\rangle—NH.R$$

The protein is treated with an alcoholic solution of FDNB and, when reaction is complete, the mixture is extracted exhaustively with ether to remove unused reagent. The yellow DNP-protein is hydrolysed with hydrochloric acid as before and the yellow DNP amino acids liberated by hydrolysis extracted into ethereal solution. The DNP amino acids in this solution will be of two types: those derived from the N-terminal group of the protein and those derived from the side reactions of FDNB with other groups in the amino acids of the protein. Included in the latter category will be the reactions of FDNB with the ε-amino group of lysine residues and the phenolic groups of tyrosine. However, provided the enzyme is pure and contains only one chain or type of chain, then there will be only one DNP amino acid derived from reaction with an α-amino group since there is only one such group free in any protein chain. The extraneous amino acid derivatives may be removed by selective washing of the ethereal extract and the material is then chromatographed on paper or more usually on glass plates covered with a thin layer of silica gel. The distance moved by the unknown DNP amino acid is compared with the movement of standard DNP amino acids and in this way the N-terminal group is identified. By colorimetric measurement of the yellow DNP-amino acid, the amount of N-terminal group derived from a given weight of the protein can be found and, knowing the molecular weight of the protein, the number of chains per molecule can be calculated. Many refinements of this technique have now been made and other similar specific reagents developed which have increased the sensitivity of the method. However, the original method developed by Sanger in the late forties is still used and its principles are used in most of the other methods.

Unfortunately the methods available for the determination of C-terminal amino acids are more cumbersome and less reliable. Some use has been made of the fact that hydrazine will react with carboxyl residues bound in peptide linkage to form the hydrazides while not reacting with free carboxyl groups. Thus if a protein is treated with hydrazine under controlled conditions the protein is split up and the hydrazides of all the amino acids except the C-terminal amino acid is formed. This free amino acid may be separated and characterized by chromatography.

To determine the amino acid sequence of a protein it is necessary to break the protein up into a limited number of smaller peptide units. This cleavage must be specific and reproducible in order that a reasonable quantity of peptides is formed for subsequent analysis. For this reason the cleavage of proteins is often carried out by the use of purified proteolytic enzymes, mostly trypsin, chymotrypsin and pepsin. It is only necessary to use very small amounts of these enzymes to hydrolyse the protein and recently active derivatives of these enzymes bound to insoluble inert polymeric supports have been produced. Using these insoluble derivatives, the enzymes can be recovered from the hydrolysis solution by centrifugation and re-used.

Consider the decapeptide:

Ala.Phe.Gly.Leu.Lys.Tyr.Val.Arg.Gly.His.

The amino acid composition can be determined and the N-terminal group can be shown to be alanine. Chymotrypsin catalyses the hydrolysis of peptide bonds where the carboxyl group is donated by an aromatic amino acid, i.e. tyrosine, phenylalanine or tryptophan. In this example treatment with chymotrypsin will give three peptides:

(A) Ala.Phe.   (B) Gly.Leu.Lys.Tyr.   (C) Val.Arg.Gly.His.

The structure of A can be confirmed by N-terminal analysis and amino acid composition. The amino acid composition of B can be determined and glycine shown to be N-terminal. The specificity of chymotrypsin makes it necessary for tyrosine to be the C-terminal amino acid residue in B. However, the sequence of the leucine and lysine cannot be found from this experiment. Similarly valine can be shown to be N-terminal in peptide C but the sequence of the arginine, glycine and histidine cannot be found. The order in which the peptides B and C are found in the original is not known although the presence of tyrosine in peptide B suggests that the order is ABC.

Trypsin catalyses the hydrolysis of peptide bonds where the carboxyl group is derived from basic amino acids such as lysine and arginine. Thus the decapeptide is hydrolysed by trypsin to give three peptides:

(D) Ala.Phe.Gly.Leu.Lys.   (E) Tyr.Val.Arg.   (F) Gly.His.

The N-terminal group of D is alanine and from the amino acid composition and the specificity of trypsin it can be inferred that lysine is C-terminal. From a comparison of D and the peptides produced from chymotrypsin hydrolysis it can be seen that the order is ABC and the sequence of peptide D is firmly established. The sequence of peptides on tryptic digestion can be similarly established as DEF and the full amino acid sequence of the decapeptide determined.

Obviously the sequence determination of a large protein molecule poses great problems but the approach is essentially the same as that detailed for the decapeptide. The protein is split into smaller well-defined peptides by the use of trypsin and chymotrypsin. These peptides are then isolated by some technique which depends on the charge of the peptide. Such techniques include electrophoresis where the charged species move in an electric field, the movement being dependent on both the charge and size of the molecule, and ion-exchange chromatography, where separation is dependent on the differential binding under different conditions to the charged ion-exchange resin. The detailed sequence of the peptides is then found and using the 'overlap' provided by the use of the two enzymes the primary sequence of the protein can be elucidated. To aid the structural determination of very large proteins certain reagents which specifically cleave at certain amino acid residues have been used, e.g. cyanogen bromide cleaves after methionine residues. This splits the protein initially into more manageable peptides. In this way the structures of proteins containing several hundred amino acids have been shown.

3.6   Active centre determination

A variety of ways have been devised for determining the nature of the amino acids at the active site of an enzyme. The physical and kinetic methods will be discussed in the next chapter. The chemical methods can be divided into two categories: those using substrate or substrate analogues and those using group specific reagents, i.e. reagents which are known to react with specific amino acid groups but which are not related to the substrate of the enzyme.

The use of diisopropylfluorophosphate for the active centre determination of proteinases is an example of the first category. Diisopropyl-fluorophosphate (DFP) reacts with the hydroxyl group of serine residues:

$$\left(\begin{array}{c}CH_3 \\ CH_3\end{array}\!\!>\!CHO\right)_2 \overset{O}{\underset{\|}{P}}.F + ROH \longrightarrow \left(\begin{array}{c}CH_3 \\ CH_3\end{array}\!\!>\!CHO\right)_2 \overset{O}{\underset{\|}{P}}.OR + HF$$

The DFP is sufficiently similar to the normal substrate to form a covalent complex with the enzyme analogous to the enzyme substrate complex. However, this complex is more stable than the substrate complex and its breakdown can be neglected. Although it is possible for DFP to react with every serine residue in the molecule, if the time and conditions of the reaction are strictly controlled, DFP reacts initially with the serine of the active site, presumably because the active site serine is more reactive and more freely accessible than other serine residues which may be tucked in some fold of secondary or tertiary structure. DFP only reacts initially with one of the two serine residues of chymotrypsin. The reacted protein is then hydrolysed and the particular amino acid residue which has reacted with the DFP determined in a manner similar to the determination of the primary sequence of a protein. If the reaction of a protein with DFP does not lead to the loss of enzymic activity then it may be assumed that serine is not one of the residues at the active site of the enzyme. There are many proteinases and esterases which are specifically inhibited by DFP and whose sequences at the active centre are known. These sequences are summarized in Table 5. The amino acids surrounding the serine residue are similar in most cases. It is believed that the mechanism of the reaction catalysed is very similar in all of these enzymes and that the substrate specificity of the enzyme is found in some other part of the active site which determines which molecules are firmly bound to the enzyme. Thus the active site of the enzyme may be subdivided into those residues responsible for binding the substrate molecule and those residues which catalyse the transformation of the bonds – binding sites and catalytic sites.

For many enzymes there is no suitable substrate or substrate analogue which forms a covalent complex and in these cases the less specific group

Table 5   Residues around the serine in the active site of some proteinases
and esterases

Chymotrypsin	Asp.Ser.Gly.
Trypsin	Asp.Ser.Gly.
Elastase	Asp.Ser.Gly.
Thrombin	Asp.Ser.Gly.
Subtilisin	Thr.Ser.Met.
Caseinase	Thr.Ser.Met.
Acetylcholine esterase	Glu.Ser.Ala.
Alkaline phosphatase	Asp.Ser.Ala.

reagents must be used. Many of the oxidoreductases rely on the sulphydryl group of cysteine at the active site for their catalytic activity. These sulphydryl groups will react with p-chloromercuribenzoate (PCMB):

$$R.SH + Cl.Hg.C_6H_4.COOH = R.S.Hg.C_6H_4.COOH + HCl$$

However, not only the SH groups at the active site will react and several other groups found in amino acid side-chains will also react. If an enzyme on reaction with PCMB loses its catalytic activity it may be impossible to distinguish between active site blocking and alteration of tertiary structure leading to the disruption of the active site conformation. This difficulty can be partly overcome by reacting the enzyme with PCMB in the presence of a large excess of substrate. The substrate will be bound to the active site and prevent its reaction with PCMB. The unreacted PCMB and substrate are now removed and the catalytic activity tested. In some cases the enzyme is still active. If this enzyme is now reacted with more PCMB in the absence of substrate, then only the active site will be left to react. If catalytic activity is now lost, then there is some group at the active centre that reacts with PCMB. If radioactively labelled PCMB is used for the second reaction than the nature and position in the protein chain of the reacting amino acid can be determined.

In a similar manner iodination of a protein may cause loss of activity if tyrosine is an active centre residue and treatment with FDNB can be used to show lysine, histidine and tyrosine residues at the active centre. None of these methods is absolutely specific and conclusive in the identification of active centre residues. However by combining methods and taking into account the evidence derived from kinetic studies, the nature of the active centre of many enzymes has been determined.

3.7   Size and shape determination

A detailed discussion of the physical techniques of determining the size and shape of enzyme molecules is outside the scope of this book but a brief description of the types of method employed is pertinent.

One of the most widely used methods of finding the approximate molecular weight of a protein is gel filtration. A solution of the protein under study is passed through a column containing small beads made of a cross-linked polymer. Small molecules such as amino acids are able to penetrate the pores of the beads while larger protein molecules are unable to penetrate and pass directly through the column, i.e. proteins are excluded from the beads. Molecules of intermediate size have more difficulty in penetrating the gel beads than the amino acids and are partially excluded. This phenomenon is shown diagrammatically in Fig. 3–4. The gel bead is chosen so that the protein under study is partially excluded.

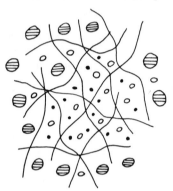

Fig. 3–4   Representation of gel filtration showing included, excluded and partially excluded molecules.

The detailed theory of this technique is extremely complex. However the practical application to the determination of molecular weight is very simple. A small volume of a concentrated solution containing proteins of known molecular weight and also the unknown protein is applied to the top of a column of gel beads, previously swollen in a suitable buffer and packed to produce a uniform column. The sample is washed through the column by the application of buffer to the top and the liquid leaving the column is collected in a fraction collector keeping the volume of each fraction as small as possible. The volume of buffer to elute a particular protein is determined by analysis of the fractions for that protein. By comparison of the elution volumes of the proteins of known molecular weight with the elution volume of the unknown protein, its molecular weight may be estimated with approximately 5% accuracy. In practice it has been found that certain proteins have an anomalous molecular weight when measured by this technique. In general these proteins are fibrous or cigar shaped and so the molecular weight does not give a true indication of their size. In the theory of gel filtration, it is assumed that proteins are behaving like spheres and do not need a specific alignment in order to enter the gel bead meshwork. Obviously this assumption is not justified for fibrous molecules and hence the molecular weight obtained is anomalous.

One of the common hydrodynamic methods used is the determination of sedimentation velocity in an ultracentrifuge. An ultracentrifuge consists of a very powerful motor which drives a rotor containing the sample under test in an evacuated chamber to minimize heating effects due to air fraction. The rotor can be spun at speeds up to 65 000 r.p.m. which develops a gravitational field in the sample of up to 300 000 times that of gravity. Before centrifugation, the proteins are distributed homogeneously in the solution but, after a period of centrifugation at a high speed, the protein molecules start to sediment as shown in Fig. 3–5. The sample nearest to

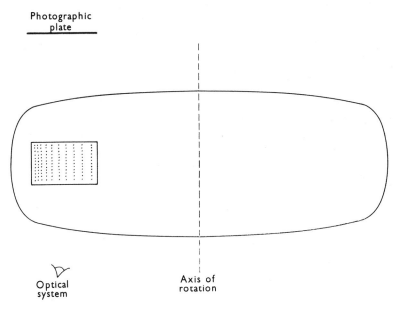

Photographic plate

Optical system

Axis of rotation

Fig. 3–5   The sedimentation of protein molecules in an ultracentrifuge.

the axis of rotation becomes less concentrated. A complex optical system is included in the ultracentrifuge which measures the rate of change of concentration with distance at varying distances from the axis of rotation. These changes are recorded automatically on a photographic plate and the rate of movement of protein in the gravitational field can be calculated. This rate is a function of both the size and shape of the protein molecule which may hence be calculated.

The use of these and similar techniques either separately or in combination has allowed the molecular weights and approximate shapes of many proteins to be determined. However, the shape is only approximate and no real indication is obtained of the spatial distributions of all the atoms by such techniques. Only X-ray diffraction analysis allows such detailed structure to be determined. X-rays are scattered by electrons and the extent of scattering is proportional to the electron density at any point in the molecule. In order that the X-ray scattering is measurable and to avoid interference of the scattering of adjacent molecules the molecules must be spatially aligned. Crystals are ordered arrangements of molecules and so a prerequisite of protein structure determination by this technique is the crystallization of the protein. This is not always easy and it may take years to grow crystals of satisfactory size and structure. Usually a metal derivative of the protein is crystallized since metals are excellent X-ray scattering centres and will provide a reference point in the molecule to which the rest of the electron density may be related. The fine detail of the diffraction photographs obtained by the scattering of X-rays by the protein crystals is analysed by high speed computers and in this way a three-dimensional picture of the electron density of the molecule is built up. Resolution of the protein structure of 0.2 nm has been obtained. As shown in Fig. 2–1 the atoms of the peptide bond are separated by approximately this distance. Atoms on adjacent parts of the protein chain may approach to within approximately 0.3 nm of each other. This gives an indication of the structural detail which may be obtained at this resolution. Unfortunately the technique is extremely expensive and laborious and this, coupled with the difficulty in obtaining satisfactory crystals, has limited its application to a relatively few proteins at the present time.

Factors affecting Enzyme Activity 4

4.1 Formation of an enzyme-substrate complex

Although, by definition, an enzyme being a catalyst is unchanged at the end of a reaction, the enzyme and substrate must have interacted in some specific physical or chemical manner during the course of the reaction. This fact may be represented in a simple form by the reaction scheme:

$$E + S \underset{k_{-1}}{\overset{k_{+1}}{\rightleftharpoons}} ES \underset{k_{-2}}{\overset{k_{+2}}{\rightleftharpoons}} E + P \tag{1}$$

where E, S and P represent enzyme, substrate and product respectively and ES is the enzyme-substrate complex formed as an intermediate. The rate constants for the formation of the complex from free enzyme and substrate and the reverse dissociation process are shown as k_{+1} and k_{-1} respectively while the rate constants for the corresponding association and dissociation reactions of enzyme and product are k_{-2} and k_{+2} respectively.

Two main types of enzyme-substrate complex can be distinguished. In some cases the complex is relatively stable and covalent bonds are formed between the enzyme and substrate, while in others the complex is transient. In the former cases it is often possible to isolate and purify the complex. Such complexes may be detected in reactions where the association between enzyme and substrate is rapid (i.e. k_{+1} is very large) but the subsequent dissociation into free enzyme and products is relatively slow, i.e. k_{+2} is very small. Chymotrypsin forms a relatively stable enzyme-substrate complex when it catalyses the hydrolysis of simple esters. In forming a complex with the ester the chymotrypsin hydrolyses the ester bond and the alcohol moiety, which is not bound to the enzyme diffuses away from the site of the reaction. The acid moiety, however, is firmly bound to the enzyme and a covalent acyl-enzyme is formed and may be isolated:

$$R_1CO.OR_2 + E \rightleftharpoons R_1CO.E + R_2OH$$

This fast reaction is followed by the slow breakdown of the acylchymotrypsin to give $R_1.COOH$ and to regenerate the free enzyme.

In most cases the complex formed is very labile and has only a transient existence. The rapid formation of the complex is followed by the equally rapid breakdown of the complex into free enzyme and products. In such cases it is never possible to isolate the complex and its formation must be followed by using extremely rapid and sensitive techniques. Complex formation with the enzyme peroxidase may be studied in this way. This enzyme catalyses the decomposition of hydrogen peroxide into water with the simultaneous oxidation of some suitable reduced species. This may be represented by the equation:

$$H_2O_2 + E \rightleftharpoons H_2O + \text{oxidized enzyme}$$

$$\text{oxidised enzyme} + H_2R \rightleftharpoons R + H_2O + E$$

The oxidized enzyme has different spectral properties from the original enzyme and although it is present in only very small amounts it may be detected with sensitive spectrophotometers.

Before proceeding to a detailed consideration of the factors affecting enzyme reactions, it is necessary to enlarge the active centre concept. It was seen in Table 5 that the region surrounding a serine in the active site of some enzymes was very similar. For these enzymes it is possible to draw up a general scheme of the mode of bond breakage. The specificity of the enzyme must lie in the nature of the groups on the enzyme which bind substrate to form the complex. These ideas have been elaborated by KOSHLAND as shown in Fig. 4-1. He considers the protein as made up of four types of amino acid residue:

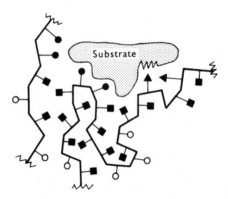

Fig. 4-1   Schematic representation of the active site. (Solid circles) binding sites; (triangles) catalytic sites acting on the substrate bond indicated by a jagged line; (open circles) non-essential residues on the surface; (squares) residues which help to maintain tertiary structure. (Reproduced with permission from KOSHLAND, 1963. *Science*, **143**, 1534.)

(a)   Non-essential. These amino acids contribute nothing directly or indirectly to the catalytic process. They may under certain circumstances be removed without loss of catalytic activity. Such residues are often found on the surface of the enzyme.

(b)   Residues involved in holding the tertiary structure of the molecule intact. Alteration of these residues although not directly concerned in binding the substrate or in its transformation will lead to a loss of catalytic activity because of the spatial disorientation of the active site.

(c) Binding residues. On close approach of the substrate to the enzyme molecule, these residues hold the substrate in position and so orientate it that the bond to be altered is brought close to the catalytic residues.

(d) Catalytic residues which actually catalyse the bond transformation.

In contrast to the old lock and key theory of enzyme action where the relationship between enzyme and substrate was exact and predetermined, KOSHLAND has put forward the theory of 'induced fit'. On binding the the substrate there is considerable freedom for the enzyme to change its conformation and to partially mould itself around the substrate so that the catalytic groups are correctly aligned. There are many modifications and variations on this idea but it serves to illustrate the molecular approach to the explanation of enzymic activity which has developed particularly during the sixties.

4.2   Enzyme concentration

The rate of an enzyme-catalyzed reaction is directly proportional to the concentration of enzyme present as shown in Fig. 4-2. The validity of this linear relationship is absolutely essential to the accurate measurement of enzyme activity. The relationship only holds true under certain rigidly defined conditions. The pH and temperature of the system must be constant and the substrate must be present in excess. If the substrate is not present in excess there will be significant variations in the substrate concentration during the course of the measurement and the extent of the variations will be dependent on the enzyme concentration. As we shall see in a later section the rate of the reaction is dependent on the substrate concentration and, when assaying an enzyme, it is necessary to ensure that

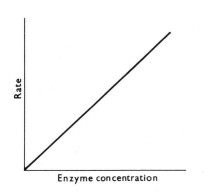

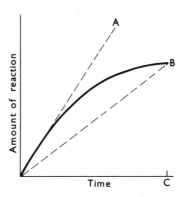

Fig. 4-2 (*left*) The effect of enzyme concentration on the rate of an enzyme reaction.

Fig. 4-3 (*right*) The time-course of an enzyme reaction, showing the difference between initial and final rate.

only an insignificant percentage of the substrate is transformed during the measurement. In general the concentration of enzyme and substrate are adjusted so that less than 1% of the substrate is used up during the measurement, the sensitivity of modern equipment allowing the accurate detection of such relatively small changes.

4.3   Time

The usual variation of the extent of reaction with time is shown in Fig. 4–3. The graph has two distinct regions. In the initial period of time the amount of substrate which has been transformed is directly proportional to the length of time for which the reaction has been proceeding. After this initial period the rate of reaction starts to decrease and the amount of reaction is no longer directly proportional to the time. Provided the substrate is present in excess the explanation of this phenomenon is the progressive loss of enzyme activity after a period of time. This may be due to the effect of heat on the tertiary structure of the enzyme or to the formation of some product or side-product of the reaction which inhibits the enzyme. In the initial investigation of any enzyme reaction it is important to establish the period over which the reaction varies linearly with time. This period can be as short as a minute while some enzymes have a linear variation over a period of days. For meaningful results the initial rate of reaction should be measured. In Fig. 4–3 the initial rate of reaction is given by the slope of the line A. If, however, the reaction had only been studied by measuring the substrate transformed after time C, then the apparent rate of reaction, given by the slope of line B, would have been considerably lower.

4.4   Substrate concentration

If the initial rate of an enzyme reaction is measured at a series of substrate concentrations, the results obtained take the form shown in Fig. 4–4. At low substrate concentrations the rate is directly proportional to the substrate concentration, while at high substrate concentrations the rate is constant. This characteristic graph is a rectangular hyperbola which has the general equation:

$$y = \frac{ax}{x+b} \qquad (2)$$

where a is the maximum value of y reached and b is the value of x when y is half the maximum value, i.e. when $y = \frac{1}{2}a$. In the specific case of the variation of rate with substrate concentration the constants a and b are given the symbols V and K_m respectively and their relationship to the graph is shown. If v is the rate at any substrate concentration, S, then they are related by the equation:

$$v = \frac{V.S}{K_m + S} \qquad (3)$$

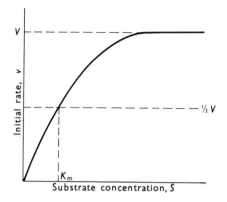

Fig. 4-4 The variation of reaction rate with substrate concentration. See text for definition of symbols.

This is the so-called Michaelis–Menten equation and over the whole range of substrate concentrations represents graphs of the type shown in Fig. 4-4. V represents the maximum rate attained by the reaction at high substrate concentrations and K_m, the Michaelis constant, is the substrate concentration when the rate is half the maximum attainable rate. Thus V and K_m are experimentally determinable constants and provided the enzyme is studied under controlled conditions, e.g. constant defined pH and temperature they are characteristic and constant for a particular enzyme.

Although the above derivation of the Michaelis–Menten equation is made from experimental results, this equation may also be predicted from theoretical considerations. It is necessary to assume that the recombination of product with enzyme to give ES complex may be neglected and that the substrate concentration is very much greater than the enzyme concentration. Under these conditions

$$V = k_{+2} . E$$

where E is the enzyme concentration

and

$$K_m = \frac{k_{-1} + k_{+2}}{k_{+1}}$$

The maximum velocity V is directly proportional to the enzyme concentration and if enzyme activity is measured in the presence of excess substrate the rate so found is a direct measure of the amount of enzyme present. The exact significance of K_m is less well-defined. It is very dependent on the validity of the assumptions made and on the applicability of the simple reaction scheme used in this derivation. In general the most satisfactory definition of the Michaelis constant is the experimental one, i.e. the substrate concentration at which the rate of the reaction is equal to half the maximum rate.

4.5   The determination of V and K_m and their significance

Although it is possible to obtain the value of V and K_m directly from the graph of rate against substrate concentration it is more usual to plot the reciprocal of the rate, $1/v$ against the reciprocal of the substrate concentration, $1/S$.

Rearranging the Michaelis–Menten equation

$$\frac{1}{v} = \frac{K_m}{V.S} + \frac{1}{V}$$

The reciprocal plot shown in Fig. 4–5 has a slope of K_m/V, the intercept on the $1/v$ axis is equal to $1/V$ and the intercept on the $1/S$ axis is equal to

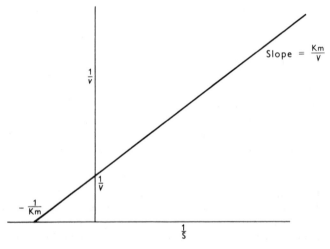

Fig. 4–5   The Lineweaver–Burk plot for the determination of K_m and V.

$-1/K_m$. In this way K_m and V can be conveniently found. K_m has the same units as the substrate concentration which is usually expressed as molarity. V has the same units as the rate, e.g. micromoles of substrate transformed per unit time. This method of expressing the results is known as the Lineweaver–Burk plot.

Since the maximum rate, V is equal to $k_{+2} . E$ if the maximum rates of two reactions are compared at equal enzyme concentrations, the relative values of the rate of breakdown of the enzyme-substrate complex can be found.

In the reaction scheme derived above and provided the assumptions are valid, K_m is equal to $(k_{-1} + k_{+2})/k_{+1}$. If it can be shown that the rate of breakdown of the ES complex to products is very slow compared to its formation from substrate, i.e. k_{-1} is very much greater than k_{+2}, then K_m

approximates to k_{-1}/k_{+1} which is the dissociation constant of the ES complex and given the symbol K_s. The reciprocal of K_s is a measure of the affinity or binding of substrate to enzyme. If the enzyme is able to act on a variety of related substrates and the above criteria are fulfilled then a comparison of the K_m values for each substrate gives an indication of the relative affinity of the enzyme for that substrate. It must be emphasized that the use of K_m as the binding constant is only justified if the rate of breakdown of ES can be shown to be very small.

4.6   pH

An enzyme is only active as a catalyst within a relatively narrow range of pH and usually the enzyme is most active at a specific pH, called the optimum pH. A typical plot of activity against pH is shown in Fig. 4–6. For most enzymes this optimum pH is near neutrality (pH 5 to pH 9) but there are some exceptional enzymes such as pepsin which act at extreme pH values.

Two explanations are possible for this phenomenon. The first considers that the enzyme is unstable at pH removed from the optimum and that at these pH values the enzyme starts to lose the tertiary structure necessary for the conformation of the active site. This process, which would, at best, be slowly reversible, may be tested by leaving a known quantity of enzyme

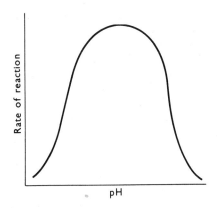

Fig. 4–6   The effect of pH on reaction rate.

at a pH removed from its optimum for a period of time and then quickly adjusting the pH back to the optimum and remeasuring the activity. If the full activity is restored, it is unlikely that the change in tertiary structure accounts for the decreased activity at the other pH. The more likely explanation in the majority of cases is that we are seeing the effects of pH on the ionization of acidic and basic groups at the active centre of the enzyme.

Consider an enzyme with an acidic group AH and basic group B. This enzyme may be represented as HA.R.B. At low pH the enzyme will exist in the form HA.R.BH$^+$. As the pH is increased the A group ionizes and the enzyme is now in the form $^-$A.R.BH$^+$. A further increase in pH leads to the formation of $^-$A.R.B. Let us assume that only the species where both A and B are charged is active in catalysis. The enzyme will start to show activity when there has been appreciable ionization of AH and the activity will progressively increase as the extent of this ionization increases. Eventually the AH will be virtually fully ionized and the enzyme will show its maximum activity. If the ionizations of AH and BH$^+$ are fairly close in their pK values, as the pH of maximum activity is reached, the BH$^+$ group will start to ionize appreciably and the activity will start to fall. There will be a progressive decrease in activity as the pH increases until the BH$^+$ is virtually completely ionized when the activity of the enzyme will become zero. This approach has been used extensively in the study of the groups responsible for enzyme activity at the active centre and in particular, if the value of V is found at each pH, it is possible to calculate the pK values of catalytic groups, as distinct from binding groups at the active centre. This provides a very useful tool in addition to the chemical methods of active centre characterization previously described.

4.7   Temperature and denaturation

Consider a reaction C + D going to E + F. The energy changes for such a reaction are shown in Fig. 4–7. The net result may be to give out heat (i.e. exothermic) or to take in heat (i.e. endothermic) but in all cases, before reaction can take place, it is necessary for the reactants to overcome the 'energy barrier' which is called the activation energy. The greater the activation energy the more difficult it is for the reaction to take place and the more heat which must be applied to the system for successful reaction. An enzyme lowers the overall activation energy of the reaction. It does so by binding C and D to itself in the orientation for reaction to take place. While the uncatalysed reaction relies on random collisions between C and D molecules the reaction chances are greatly enhanced by binding in the catalysed reaction. The energy changes for the catalysed reaction are shown in Fig. 4–8, where it can be seen that a small activation energy is needed for the formation of the enzyme-substrate complex and also for the dissociation of the enzyme-product complex and the activation energy for the overall reaction is greatly decreased.

Consider the effect of temperature on an enzyme-catalysed reaction. As the temperature is increased so the inherent energy of the system is increased and more molecules obtain the necessary activation energy for reaction to take place. The rate of reaction therefore increases with temperature. If there were no other factors to take into consideration, this increase in reaction rate with temperature would continue ad infinitum. However, as the temperature is increased above a certain point the energy of the system is sufficient to cause the breakdown of hydrogen bonding and

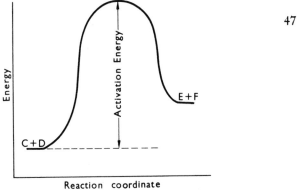

Fig. 4-7 Energy changes during the course of the reaction $C + D = E + F$.

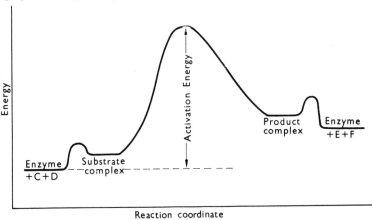

Fig. 4-8 Energy changes as in Fig. 4-7 for the enzyme catalysed reaction.

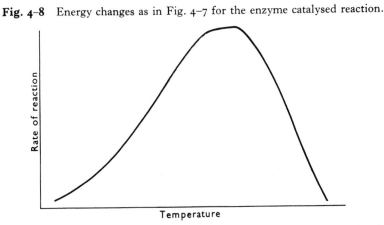

Fig. 4-9 The influence of temperature on the rate of an enzyme-catalysed reaction.

many other forces which hold the tertiary structure. The enzyme starts to lose its activity and eventually becomes completely inactive, a process known as denaturation. This behaviour is shown in Fig. 4–9. At low temperature the activity is steadily increasing but, in the region of 40°C for most enzymes, the tertiary structure starts to alter and there comes a point where the increase of reaction due to temperature on the activation of molecules is equal to the decrease of reaction rate due to destruction of tertiary structure. At this point the activity is a maximum and this temperature is often known as the optimum temperature.

4.8 Irreversible and reversible inhibitors

The use of group specific reagents such PCMB, DFP and FDNB is an example of irreversible inhibition. Once the reagent has formed a covalent bond with the enzyme, it is relatively difficult to remove the group and restore the enzyme activity. When using reversible inhibitors, however, the inhibitor is rarely covalently bound and may be easily removed with the restoration of enzyme activity by techniques such as dialysis, dilution or addition of substrate in excess.

Reversible inhibitors can be divided into two main categories: competitive and non-competitive and allosteric. We shall leave a discussion of the allosteric inhibitors to the next chapter and concentrate here on competitive and non-competitive. Competitive inhibitors, as the name implies, compete with substrate for binding sites on the enzyme molecule and so one might expect will have structural similarities to the substrate. Non-competitive inhibitors on the other hand attack the catalytic sites on the enzyme and, although the enzyme is still able to bind substrate, the catalysis (i.e. the bond transformation) is unable to take place. From this it can be seen that the nature of inhibitors will be able to give help in the elucidation of the binding and catalytic sites at the active centre.

The easiest way to differentiate in practice between competitive and non-competitive is to determine the variation of the rate of a reaction with substrate concentration at a series of inhibitor concentrations. Figure 4–10 shows typical results obtained with a competitive inhibitor. When $1/v$ is plotted against $1/S$ at the various inhibitor concentrations a series of straight lines is obtained which intersect on the $1/v$ axis. This point corresponds to the maximum velocity V. The lines intersect at various points on the $1/S$ axis and so the apparent value of the Michaelis constant changes in the presence of different concentrations of inhibitor. These results are to be expected from the theory of action of competitive inhibitors, since they affect the binding of substrate and hence K_m but, once the substrate is bound, there is no effect on the subsequent catalysis and hence on V.

Figure 4–11 shows comparable results for a non-competitive inhibitor. The lines intersect at the $1/S$ axis giving the same value of K_m in each case. This is also to be expected since the inhibitor does not interfere with the

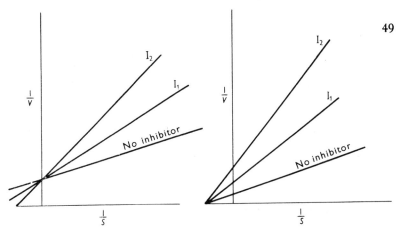

Fig. 4-10 Reciprocal plots for a competitive inhibitor showing intersection on the $1/v$ axis.

Fig. 4-11 Reciprocal plots for a non-competitive inhibitor showing intersection on the $1/S$ axis.

binding of substrate. The inhibitor acts by blocking catalytic groups and hence the value of V as obtained from the intercept on the $1/v$ axis is different in each case.

A typical example of competitive inhibition is the effect of malonate on the enzyme succinate dehydrogenase. The enzyme catalyses the conversion of succinate to fumarate:

$$\begin{array}{ccc}
\text{CH}_2\text{COOH} & \text{CHCOOH} & \diagup\text{COOH} \\
| & || & \text{CH}_2 \\
\text{CH}_2\text{COOH} & \text{CHCOOH} & \diagdown\text{COOH} \\
\text{Succinate} & \text{Fumarate} & \text{Malonate}
\end{array}$$

The structure of malonate is very similar to that of succinate and hence the enzyme is able to bind easily. However, the two hydrogens that are removed from succinate are not present in malonate and hence the enzyme cannot catalyse the dehydrogenation of this inhibitor. Acetate is neither an inhibitor nor substrate of this reaction. At physiological pH both of the carboxyl groups of the succinate are ionized and these facts, taken as a whole, suggest that there are two positively charged groups on the enzyme which are responsible for binding the substrate. Competitive inhibitors are particularly useful since they are usually specific to the particular enzyme. They may be used to selectively block some area of metabolism and hence one finds that many drugs are based on the competitive inhibition of key enzymes in metabolism.

The inhibition of arylsulphatase by cyanide is an example of non-competitive inhibition. The inhibitor has no structural similarity to the substrate and is relatively non-specific. Cyanide is known to inhibit many enzymes and can be shown to act by combining with metallic ions which serve as prosthetic groups in many cases.

Having considered in some detail the structure and properties of enzymes, it is pertinent to examine present theories of the mechanism of action of a few specific enzymes. This in turn leads to a consideration of the influence of enzymes on individual reactions and the role of enzymes in the integration and control of metabolism.

5.1 Mechanism of enzyme action

Each enzyme-catalysed reaction will obviously have a different mechanism from any other. However, from our present knowledge of the few enzyme mechanisms which have been investigated in detail, certain basic general ideas are starting to emerge. The universal applicability of such ideas is open to conjecture.

5.1.1 *Ribonuclease*

In a preliminary consideration of the mechanism of action of this enzyme (see Fig. 1–3) the catalytic activity was postulated to arise by the accurate spatial orientation of the active groups near the bond in RNA to be hydrolysed. Among the residues at the active centre of ribonuclease are two lysines and two histidines. The lysines are positively charged at pH near to neutrality (the optimum pH of the enzyme) and are used to bind the RNA, which has many negatively charged ionized phosphate groups, to the enzyme in the correct orientation. The histidines are the catalytic groups. The imidazole residue of histidine ionizes in the pH range 5·5 to 7 and the actual pK is very dependent on the micro-environment. In Fig. 1–3 groups A and B are both histidine and at the start of the reaction A is mostly ionized while B is unionized. During the formation of cyclic phosphate the group B loses its proton while A gains a proton from the RNA. In the course of the hydrolysis of the cyclic phosphate the process is reversed. The ability of histidines to act in this way is related to the fact that their pK is very near neutrality and the histidine finds it easy to function as both an acid and base in this pH region. The state of ionization of the imidazole group in these systems is very dependent on the local concentrations of hydrogen ions and the nature of charged groups surrounding the residue.

There are now reports in the literature of a considerable number of enzymes which have two histidines in their active centre and where this concerted acid-base mechanism is believed to operate. This finding is not altogether surprising because many organic reactions are catalysed by both acid and base individually and hence it is to be expected that the combined action of acid and base acting in concert would prove an extremely effective catalyst. Only by fixing the positions of the acid and base relative to each other and by spatially orientating with respect to the bond that is to be broken can this concerted action of the protein be achieved. There have been several successful attempts to synthesize low molecular weight compounds where these criteria are fulfilled. These synthetic enzyme analogues show catalytic activity and, although in most cases they are inferior in their power to natural enzymes, the fact that they have activity is further evidence of the validity of the postulated mechanisms.

5.1.2 'Serine' proteinases

In a previous discussion of the active centre of this group of enzymes it was shown that in all probability the mechanism of catalysis was similar in each case and that the binding sites of the enzyme conferred specificity on the individual enzymes of the group. It has also been found that a histidine is present at the active centre of all the enzymes in this group in addition to the serine previously mentioned. The mechanism of hydrolysis of simple esters by chymotrypsin is very similar to the catalysis of RNA hydrolysis by ribonuclease. Once again there is a concerted acid-base attack. In the case of ribonuclease, after the rupture of O—P ester bond, the free acid function liberated is esterified internally to give the cyclic phosphate. No possibility of such internal ester formation exists in the case of the ester hydrolysis catalysed by chymotrypsin. The acid in this case, which is liberated on hydrolysis of the ester bond, is esterified to the serine of the enzyme and the second phase of the mechanism is the hydrolysis of this to regenerate the free enzyme in a manner analogous to the hydrolysis of the cyclic phosphate. Thus the major difference in the mechanism of ribonuclease and chymotrypsin is that in the former case there is internal ester formation while in the latter the ester formation is with the enzyme.

It is now believed that the specificity of the various serine proteinases lies not in the nature of the active binding residues but in the residues surrounding the binding sites. For example, near the binding site of chymotrypsin there is believed to be a non-polar hydrocarbon cleft which can easily accommodate the aromatic residues which are part of the specificity requirements of chymotrypsin. This cleft, by its non-polar nature would be unable to accommodate the charged basic groups inherent in the specificity of trypsin and thus the residues around the binding site of trypsin must be very different and conducive to the lysine and arginine residues.

5.3.1 Glyceraldehyde-3-phosphate dehydrogenase

The systematic name of this enzyme, glyceraldehyde-3-phosphate: NAD oxidoreductase is not completely informative in this case, since although the reaction does involve the oxidation of an aldehyde to an acid,

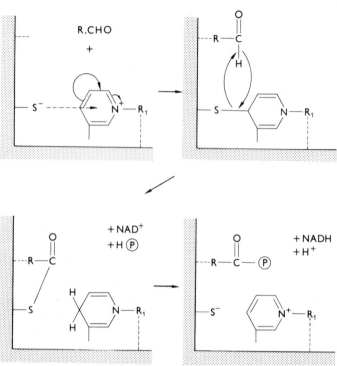

Fig. 5–1 A posulated mechanism for glyceraldehyde-3-phosphate dehydrogenase. A detailed explanation of the mechanism is given in the text. For clarity only the aldehyde portion of the substrate has been included, only the nicotinamide ring of the coenzyme and phosphoric acid has been represented as H(P).

the free acid is not found but only its phosphate ester. The overall reaction is the conversion of glyceraldehyde-3-phosphate to 1.3 diphosphoglyceric acid and represented by the following equation:

$$\text{glyceraldehyde-3-phosphate} + \text{NAD}^+ + \text{phosphate}^- \rightleftharpoons \text{1.3diphosphoglyceric}$$
$$\text{acid} + \text{NADH}$$

A tentative mechanism of the reaction is shown in Fig. 5–1. The ionized sulphydryl group is negatively charged and hence nucleophilic. It is electrostatically bound to the essentially electrophilic nicotinamide ring of the NAD which accounts for the reasonably strong binding of the coenzyme to the enzyme. The coenzyme cannot be removed by repeated dialysis or recrystallization of the enzyme and more drastic techniques such as absorption on activated charcoal are necessary for complete removal. The glyceraldehyde-3-phosphate is bound through its negatively charged phosphate group to some positively charged amino acid residue near the cysteine at the active centre. The first stage in the catalysis is the removal of hydrogen from the substrate and transfer to the NAD. The reduced NAD, NADH, is not electrophilic and this releases the ionized sulphydryl group which becomes free to accept the newly formed acyl group of the substrate forming an acyl-enzyme. The second stage of the reaction is the hydrolysis of the acyl-enzyme by phosphoric acid forming 1.3.diphosphoglyceric acid and regenerating the free enzyme. The NADH is then replaced by NAD from the coenzyme pool and the process is ready to be repeated. Here again, as in the case of chymotrypsin, an acyl-enzyme intermediate is formed but this time cysteine rather than serine is the residue involved.

5.2 Quaternary structure, protein sub-units and allosteric effects

Although we have been considering the structure of enzymes as simple single protein chains, this picture is a gross oversimplification of many enzyme structures. In the case of glyceraldehyde-3-phosphate dehydrogenase, the molecular weight normally obtained is approximately 140 000 and there have been shown to be between 14 and 18 cysteine residues in the molecule. However, in the presence of ATP the molecule dissociates into four essentially identical sub-units of molecular weight 36 000 and each having four cysteine residues.

Whenever this type of sub-unit pattern is observed, it is common to find that the enzyme is sensitive to regulation by substrate and many other small molecules and that it loses its catalytic activity when dissociated into sub-units. Consider the case where a substrate is reacting with a dimeric enzyme. There are two possibilities: the two sub-units may act independently and function as two distinct active centres or the binding of enzyme to one site may influence the binding of the substrate to the other site. In general it is found that the binding of substrate to one sub-unit increases the affinity of the other sub-unit for substrate and thus facilitates the binding of the second substrate molecule. A schematic representation of this phenomenon is shown in Fig. 5–2a. The first substrate molecule fits into the wedge on sub-unit A and in doing so induces a conformational change in the sub-unit causing the wedge to close around the substrate. The other sub-unit which is intimately bound to sub-unit A is also affected by this induced fit and in this case the wedge is opened thereby allowing easier access of the second substrate molecule to sub-unit B and facilitating its binding. This representation is a gross oversimplification of the mechanism, which is referred to as cooperativity of substrate binding, but serves to illustrate the type of phenomenon which is occurring.

The whole range of effects which may be obtained when molecules are bound to an enzyme at a site distant from the active site of the enzyme and yet drastically influence the catalytic acitivity of that active site are referred to as allosteric effects. By no means all such binding leads to increased

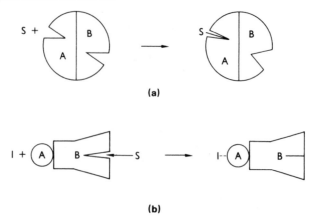

(a)

(b)

Fig. 5-2   Allosteric effects. (a) cooperativity of substrate binding. (b) allosteric inhibition. The representation is purely diagrammatic.

activity. In some cases the binding of a molecule to a site other than the active centre, called an allosteric site, may decrease the catalytic activity and and such molecules are known as allosteric inhibitors. Converse the binding to the allosteric site may lead to increased catalytic activity, i.e. allosteric activation. A schematic representation of allosteric inhibition is shown in Fig. 5–2b. The binding of the inhibitor molecule to the allosteric site causes a conformational change in sub-unit A, which is transmitted via the binding of the two sub-units to sub-unit B, which in consequence loses its ability to bind substrate and hence to catalyse the reaction.

The two sub-units in Fig. 5–2b have been represented as different. It is often found that the functional enzyme molecule, i.e. the enzyme found acting in the cell, is composed of two types of sub-unit. One sub-unit referred to as the catalytic sub-unit is responsible for the activity of the enzyme while the other sub-unit, the regulatory sub-unit, serves to provide the allosteric site and hence to allow modification of the enzyme activity. In certain cases it is possible to dissociate the sub-units and to reassemble an enzyme molecule containing only catalytic sub-units and thus freed from allosteric control.

In the same way that the binding site of an enzyme has a specificity for substrate, the allosteric sites are also specific in their binding of effector molecules. This specificity allows regulation of enzyme activity by specific molecules that bear no resemblance to substrate, this regulation being extremely important in the control and integration of metabolism. An important example of such a control system is the enzyme, phospho-fructokinase. This enzyme catalyses the phosphorylation of a sugar phosphate, fructose-6-phosphate according to the equation:

Fructose-6-phosphate + ATP → Fructose–1.6-diphosphate + ADP

This enzyme is essentially irreversible, a separate enzyme fructose-1.6-diphosphatase catalysing the hydrolysis of the diphosphate to the monophosphate. This reaction lies on the pathway of conversion of carbohydrates to pyruvic acid, i.e. the glycolytic pathway and the pyruvic acid is subsequently oxidised to CO_2 with the production of a large amount of energy which is stored as ATP. Although ATP is a substrate for the enzyme, inhibition occurs at relatively high ATP concentrations and the enzyme disaggregates. It has been suggested that the enzyme has a regulatory site which binds ATP when the concentration of ATP reaches a critical threshold value and disaggregation results. This prevents the catalysis and so effectively stops the continued oxidation of carbohydrates. The build up of excessive stores of ATP is thus prevented and the cell·retains its potential energy in the form of carbohydrate. There are many other examples of this type where the allosteric nature of the enzymes allow the subtle control of metabolism so that the cell can channel foodstuffs to the best usage at any point in time.

In summary, enzymes whose activity is subject to allosteric control may be disaggregated into inactive sub-units, have an allosteric site in addition to an active centre and often show cooperativity in their binding of substrate molecules.

5.3   Lactose synthetase

Some enzymes have been characterized where the binding of a sub-unit alters the specificity of the binding site and hence the reaction catalysed. A novel example of this type of change is shown by lactose synthetase. The membrane of the Golgi complex contains an enzyme, acetyllactosamine synthetase which catalyses the synthesis of a substituted disaccharide, acetyl lactosamine, from a nucleotide derivative of galactose, UDP galactose, and an acetylated amino derivative of glucose, acetylglucosamine:

$$\text{UDP galactose} + \text{acetylglucosamine} \rightleftharpoons \text{Acetyllactosamine} + \text{UDP}$$

Acetyllactosamine synthetase combines specifically with a milk protein, lactalbumin which has not been shown to possess any catalytic activity. The free enzyme is four times more active in catalysing the production of acetyllactosamine than the complex with the lactalbumin.

To a lesser extent acetyllactosamine synthetase is able to catalyse the synthesis of the disaccharide, lactose, itself:

$$\text{UDP galactose} + \text{glucose} \rightleftharpoons \text{Lactose} + \text{UDP}$$

The complex with lactalbumin is approximately 70 times more effective in catalysing the production of lactose than the free enzyme. This switch has important biological implications.

Acetyllactosamine synthetase is always present on the Golgi membrane and is responsible for the addition of sugar residues to the membrane.

Lactalbumin is not normally present. When lactation is about to be initiated, increased protein synthesis takes place on the ribosomes of the endoplasmic reticulum of mammary gland, mediated by the action of hormones. Among the proteins produced is lactalbumin which travels from the endoplasmic reticulum to the Golgi along the cisternal space. On meeting a molecule of acetyllactosamine, the molecule of lactalbumin combines with it and switches the enzyme from the production of ace-tylated amino sugar for membrane incorporation to the production of lactose, which is secreted by the mammary gland and is the chief carbo-hydrate of milk. Lactalbumin, which is only loosely bound by the enzyme, is secreted also in the milk and the production of lactose, and hence milk, is dependent on the continued production of lactalbumin, which in its turn contributes to the protein content of the milk. When milk production ceases, the enzyme, which has remained firmly bound to the membrane, continues with its original role in the development of membrane.

5.4 Feedback inhibition and the control of metabolism

Although throughout this book the emphasis has been on individual enzymes, their structures and their properties, the ultimate aim must always be the understanding of the role of enzymes in the metabolism of the cell. Most enzymes catalyse one of a series of related reactions and it is the overall result of the series of reactions that is important rather than the individual reaction.

Consider a sequence A→B→C→D→E→F. It is often found that the end product of a reaction sequence inhibits the enzyme responsible for catalysing the first reaction in the sequence. In this case F inhibits the enzyme catalysing the conversion of A to B. This phenomenon is known as negative feedback inhibition. There is a threshold value of the concentra-tion of F below which inhibition can be neglected and over which inhibition progressively increases. In this way the overproduction of F can be avoided.

A well documented example of feedback inhibition is seen in the pro-duction of isoleucine from threonine. The first step in the process is the oxidative deamination of threonine by the enzyme threonine deaminase to give α-ketobutyrate which is then converted to the isoleucine by a complex series of reactions:

$$\text{Threonine} \rightarrow \alpha\text{-ketobutyrate} \rightarrow \rightarrow \rightarrow \rightarrow \rightarrow \rightarrow \text{Isoleucine}$$

Threonine deaminase is specifically inhibited by high concentrations of isoleucine and this inhibition is allosteric. At first sight, it might appear that isoleucine could act as a competitive inhibitor of threonine deaminase because of its structural similarity to threonine. However, other amino acids do not inhibit this enzyme and the inhibition by isoleucine has been shown to be allosteric. By this mechanism the overproduction of isoleucide can be avoided and the threonine channelled into the synthesis of other amino acids.

5.5   Hormonal control of metabolism—glycogen breakdown

Many hormones exert their influence through a direct or indirect allosteric effect on key metabolic enzymes. An example of this is the mobilization of sugar reserves in response to stress or low circulating blood sugar levels, mediated by the hormones adrenalin and glucagon respectively. Dietary carbohydrate in excess of the immediate needs of the body is stored, particularly in liver and muscle, in the form of a complex branched polysaccharide, glycogen. In times of stress the energy requirements of the body are increased and this increase is mainly satisfied by increased oxidation of carbohydrate particularly glycogen. Adrenalin is the hormone secreted during stress. Similarly, under more normal circumstances, there is a need to keep the level of glucose circulating in the blood within narrow limits and if this level falls, glucagon is secreted in an attempt to raise the glucose level.

Glycogen phosphorylase catalyses the breakdown of glycogen to glucose 1-phosphate:

$$\text{Glycogen} + \text{phosphate} \rightleftharpoons \text{glucose-1-phosphate}$$

Glycogen phosphorylase exists in two forms; the 'a' form is always active and in muscle has a molecular weight of 500 000, the 'b' form is usually inactive and has a molecular weight of 250 000. The two forms are interconvertible and the conversion of b to a is catalysed by phosphorylase kinase. This kinase also exists in two forms and the conversion of its inactive to active form requires cyclic AMP, which is a cyclic phosphate derived from AMP and structurally similar to the intermediate cyclic phosphate formed during ribonuclease action. The adrenalin and glucagon released into the circulation bind to specific sites on the surface of a cell and in so doing increase the level of cyclic AMP within the cell. This in turn increases the active kinase which increases the active phosphorylase 'a' giving glycogen breakdown and increased circulating sugar levels. The system is summarized in Fig. 5–3.

Cyclic AMP has been similarly implicated in other hormonally controlled systems and is sometimes known as the second messenger. The hormone, the first messenger, reaches a target cell and releases within that cell the cyclic AMP which has a direct allosteric effect on some key enzyme. However, by no means all hormonal influence is mediated through cyclic AMP and there are cases where the hormone has a direct allosteric effect on the key enzyme.

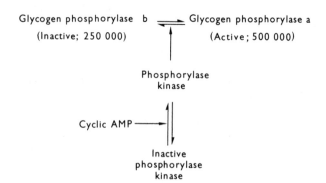

Fig. 5-3 The glycogen phosphorylase system and its hormonal control.

5.6 Cellular organization of enzymes and multienzyme systems

Of over a thousand enzymes known, it is possible to find up to five hundred in any single cell, although not all necessarily active at the same time. It is obvious that chaos would result unless there were some organization within the cell. There are various sub-cellular particles and these have specific functions and hence contain specific enzymes. For example, the mitochondrion contains the enzymes of the tricarboxylic acid cycle and the electron transport chain. It is thus the main site of energy production in the cell. Similarly the nucleus contains the enzymes responsible for nucleic acid metabolism. By this compartmentalization the enzymes involved in specific areas of metabolism are brought together and the smooth flow of processes ensured.

Metabolic pathways may be roughly divided into two types. In the first type there is a branched sequence of reactions which lead to many end products. Many catabolic (i.e. degradative pathways) are of this type. Consider the oxidation of carbohydrates by the glycolytic pathway. The beginning of this pathway is diffuse. A variety of carbohydrates including glucose, fructose and glycogen can act as starting material. The pathway, as usually written, shows the conversion of these materials to pyruvate and hence to acetyl coenzyme A. There are many branches on the pathway leading to the formation of other essential cellular constituents such as glycerol and amino acids. The fate of a molecule of glucose entering the pathway is very dependent on the metabolic state of the cell. If the cell needs energy it will be converted to pyruvate, if the cell is actively synthesizing protein it will go to amino acid and so on. The control of the route of the glucose molecule will be the activities of the enzymes at the branch points and these, in turn, will be influenced by factors such as feedback inhibition. In this way high concentrations of ATP in the cell will allow the glucose to be converted to some amino acid since ATP concentration is a measure of the available energy of the cell. The enzymes of all branched pathways are soluble and so allow the free movement of substrate from one enzyme in the sequence to another. At a branch point the controlling factor will be the relative enzyme activities.

In the second type there is a linear sequence of reactions which leads to a specific end product. In these cases the enzymes of a pathway are often found associated together in a large, relatively insoluble, multienzyme complex. A good example is the pathway for the biosynthesis of fatty acids. The overall process may be thought of as the production of a fatty acid containing four carbon atoms from the joining of two molecules each containing two carbon atoms. This is an over-simplification since one of the 2C units is converted to 3C which then combines with a 2C to give a 5C molecule. This loses carbon dioxide to become a 4C molecule which is reduced in two stages to form the saturated fatty acid. This process is repeated until a suitable chain length is built up. As can be seen in Fig. 5–4, seven enzymes are involved in the process and a diagrammatic representation of the 'fatty acid synthetase' shows correspondingly seven sub-units. If these are in the correct order, then a molecule of substrate may be passed from one enzyme to the next in a controlled and spatially advantageous manner. The anabolic processes of the cell are often of this type.

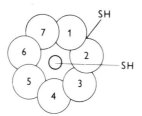

Fig. 5–4   The fatty acid 'synthetase' showing the type of reaction catalysed by each protein unit. 1. Transfer of 3C to central SH. 2. Transfer of 2C to peripheral SH. 3. Condensation. 4. First reduction. 5. Dehydration. 6. Second reduction. 7. Transfer of product to sub-unit 2.

The control and integration of metabolism is dependent on the properties of the individual enzymes, their sub-cellular localization and the physical orientation of the enzymes within those particles. This intricate and extremely complex machinery allows the cell to function smoothly and to deal within most internal and external stimuli.

Further Reading

BERNHARD, S. A. (1968). *The structure and function of enzymes*. W. A. Benjamin, New York.

CHANGEUX, J. P. (1965). *The control of biochemical reactions*. Scientific American Offprint No. 1008. Freeman, K., San Francisco.

COHEN, G. (1968). *The regulation of cell metabolism*. Holt, Rinehart and Winston, New York.

DIXON, M. and WEBB, E. C. (1964). *Enzymes*. Longmans, London.

HIRS, C. H. W. (ed.) (1967). Methods in Enzymology, Volume 10, *Enzyme structure*. Academic Press, New York.

KENDREW, J. C. (1961). *The three-dimensional structure of a protein molecule*. Scientific American Offprint No. 121. Freeman, San Francisco.

NEURATH, H. (1964). *Protein-digesting enzymes*. Scientific American Offprint No. 198. Freeman, San Francisco.

PHILLIPS, D. C. (1966). *The three-dimensional structure of an enzyme*. Scientific American Offprint No. 1055. Freeman, San Francisco.

SETLOW, R. B. and POLLARD, E. C. (1962). *Molecular Biophysics*. Addison-Wesley, London.

STEIN, W. H. and MOORE, S. (1961). *The chemical structure of proteins*. Scientific American Offprint No. 80, Freeman, San Francisco.

WESTLEY, J. (1969). *Enzymic Catalysis*. Harper and Row, London.